Embedded Systems

Series Editors
Nikil D. Dutt
Peter Marwedel
Grant Martin

For further volumes:
http://www.springer.com/series/8563

Andreas Hansson · Kees Goossens

On-Chip Interconnect with Aelite

Composable and Predictable Systems

Springer

Andreas Hansson
Research & Development ARM Ltd.
Cambridge, United Kingdom
andreas.hansson@arm.com

Kees Goossens
Eindhoven University of Technology
Eindhoven, The Netherlands
k.g.w.goossens@tue.nl

ISBN 978-1-4419-6496-0 e-ISBN 978-1-4419-6865-4
DOI 10.1007/978-1-4419-6865-4
Springer New York Dordrecht Heidelberg London

Library of Congress Control Number: 2010937102

Printed on acid-free paper

Springer is part of Springer Science+Business Media (www.springer.com)

Contents

Chapter 1
Introduction

Embedded systems are rapidly growing in numbers and importance as we crowd our living rooms with digital televisions, game consoles and set-top boxes and our pockets (or maybe handbags) with mobile phones, digital cameras and personal digital assistants. Even traditional PC and IT companies are making an effort to enter the *consumer-electronics* business [5] with a mobile phone market that is four times larger than the PC market (1.12 billion compared to 271 million PCs and laptops in 2007) [177]. Embedded systems routinely offer a rich set of features, do so at a unit price of a few US dollars, and have an energy consumption low enough to keep portable devices alive for days. To achieve these goals, all components of the system are integrated on a single circuit, a *System on Chip* (SoC). As we shall see, one of the critical parts in such a SoC, and the focus of this work, is the on-chip interconnect that enables different components to communicate with each other.

In this chapter, we start by looking at trends in the design and implementation of SoCs in Section 1.1. We also introduce our example system that serves to demonstrate the trends and is the running example throughout this work. This is followed by an overview of the key requirements in Section 1.2. Finally, Section 1.3 lists the key components of our proposed solution and Section 1.4 provides an overview of the remaining chapters.

1.1 Trends

SoCs grow in complexity as an increasing number of independent *applications* are integrated on a single chip [9, 50, 55, 146, 177]. In the area of portable consumer systems, such as mobile phones, the number of applications doubles roughly every 2 years, and the introduction of new technology solutions is increasingly driven by applications [80, 88]. With increasing application heterogeneity, system-level constraints become increasingly complex and *application requirements*, as discussed next, become more multifaceted [152].

1.1.1 Application Requirements

Applications can be broadly classified into *control-oriented* and *signal-processing* (streaming) applications. For the former, the *reaction time* is often critical [144].

A. Hansson, K. Goossens, *On-Chip Interconnect with Aelite*, Embedded Systems,
DOI 10.1007/978-1-4419-6865-4_1, © Springer Science+Business Media, LLC 2011

Performance gains mainly come from higher clock rates, more deeply pipelined architectures and instruction-level parallelism. Control-oriented applications fall outside the scope of this work and are not discussed further. Signal-processing applications often have *real-time requirements* related to *user perception* [144], e.g. video and audio codecs, or requirements dictated by *standards* like DVB, DAB and UMTS [87, 131]. For signal-processing applications, an increasing amount of data must be processed due to growing data sets, i.e. higher video resolutions, and increasing work for the data sets, i.e. more elaborate and computationally intensive coding schemes [144]. As a result, the required processing power is expected to increase by 1000 times in the next 10 years [21] and the gap between the processing requirement and the available processing performance of a single processor is growing super-linearly [88].

Delivering a certain performance is, however, not enough. It must also be performed in a timely manner. The individual applications have different real-time requirements [33]. For *firm real-time* applications, e.g. a Software-Defined Radio [131] or the audio post-processing filter, illustrated in Fig. 1.1, deadline misses are highly undesirable. This is typically due to standardisation, e.g. upper bounds on the response latency in the aforementioned wireless standards [87], or perception, e.g. steep quality reduction in the case of misses. Note that firm real-time only differs from *hard real-time*, a term widely used in the automotive and aerospace domain, in that it does not involve safety aspects. *Soft real-time* applications, e.g. a video decoder, can tolerate occasional deadline misses with only a modest quality degradation. In addition, *non-real-time* applications have no requirements on their temporal behaviour, and must only be functionally correct.

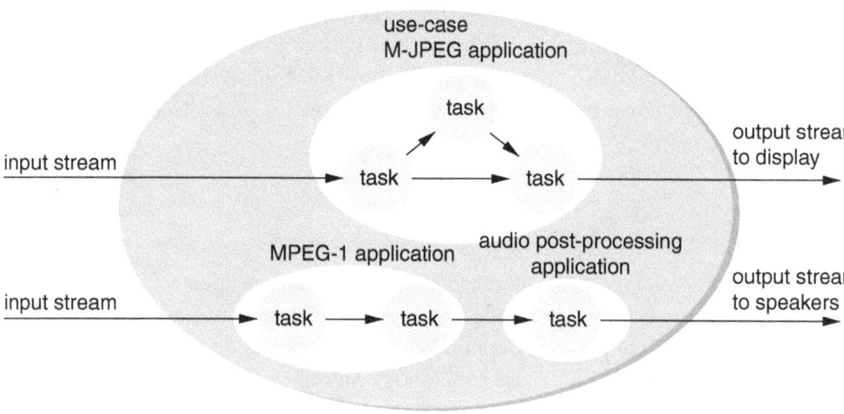

Fig. 1.1 Application model

Each application has its own set of requirements, but the SoC typically executes many applications *concurrently*, as exemplified by Fig. 1.1. Furthermore, applications are started and stopped at run time by the user, thus creating many different *use-cases*, i.e. combinations of concurrent applications [72, 138]. The number of use-cases grows roughly exponentially in the number of applications, and for every use-case, the requirements of the individual applications must be fulfilled. Moreover, applications often span multiple use-cases and should not have their requirements violated when other applications are started or stopped [72, 156]. Going from the trends in application requirements, we now continue by looking at how the *implementation and design* of SoCs is affected.

1.1.2 Implementation and Design

As exemplified in Fig. 1.1, applications are often split into multiple *tasks* running concurrently, either to improve the power dissipation [166] or to exploit task-level parallelism to meet real-time requirements that supersede what can be provided by a single processor [172]. The tasks are realised by hardware and software *Intellectual Property* (IP), e.g. accelerators, processors and application code. For an optimum balance between performance, power consumption, flexibility and efficiency, a *heterogeneous mix* of processing elements that can be tuned to the application domain of the system is required [21]. This leads to systems with a combination of general-purpose processors, digital-signal processors, application-specific processors and dedicated hardware for static parts [144]. Different IP components (hardware and software) in the same system are often developed by *unrelated design teams* [89], either in-house or by independent IP vendors [51]. The diversity in origin and requirements lead to applications using a diverse set of programming models and communication paradigms [117].

The rising number of applications and growing need for processing power lead to an increased demand for hardware and software. Figure 1.2 shows the well-known hardware design gap, with capability of technology doubling every 18 months (+60% annually), and hardware productivity growing more slowly, doubling every 24 months (+40% annually). The hardware design productivity relies heavily on *reuse* of IP and *platform-based design* [88], and improved over the last couple of years by filling the silicon with regular hardware structures, e.g. memory. Still,

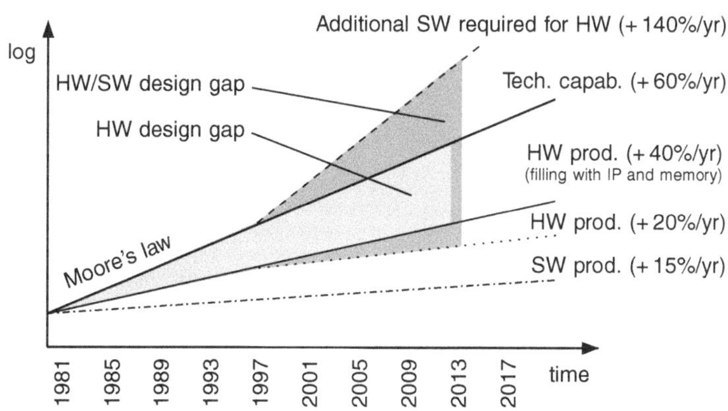

Fig. 1.2 Hardware (HW) and software (SW) design gaps [89]

systems are not fully exploiting the number of transistors per chip possible with today's technology, and the gap continues to grow [80].

The increasing number of transistors on a SoC offers more integration possibilities, but the diminishing feature size complicates modularity and scalability at the physical level [103]. Global synchronisation is becoming prohibitively costly, due to process variability and power dissipation, and the distribution of low-skew clock signals already accounts for a considerable share of power consumption and die area [89, 139]. Moreover, with a growing chip size and diminishing feature size, signal delay is becoming a larger fraction of the clock-cycle time [35], and cross-chip signalling can no longer be achieved in a single clock cycle [89, 162].

In addition to the consequences at the physical level, the growing number of IPs also has large effects at the architectural level. The introduction of more IP promises more parallelism in computational and storage resources. This, however, places additional requirements on the resources involved in communication that have to offer more parallelism as well.

While the hardware design gap is an important issue, the hardware/software design gap is far more alarming. This gap is quickly growing as *the demand for software* is currently doubling every 10 months (+140%/year), with recent SoC designs featuring more than a million lines of code [80]. The large amount of software is a response to evolving standards and changing market requirements, requiring *flexible and thus programmable platforms*. Despite much reuse of software too [189], the productivity for hardware-dependent software lags far behind hardware productivity and only doubles every 5 years (+15%/year).

A dominant part of their overall design time is spent in verification, thus limiting productivity. New designs greatly complicate the verification process by increasing the level of concurrency and by introducing additional application requirements [89]. Already, many of the bugs that elude verification relate to timing and concurrency issues [189]. The problem is further worsened by a growing sharing of resources. As a result of the sharing, applications cannot be verified in isolation, but must first be integrated and then verified together. The *monolithic* analysis leads to an explosion of the behaviours to cover, with negative impact on verification time [166], whether done by simulation of formal analysis. Furthermore, the dependencies between applications severely complicate the protection and concurrent engineering of IP, which negatively affects the *time and cost* of design.

1.1.3 Time and Cost

In addition to the problems related to the ability to design current and future systems, these systems must also be designed with a low cost and a low *Time To Market (TTM)*, as illustrated in Fig. 1.3. Portable consumer SoCs have *as-soon-as-possible requirements* on TTM [89]. The requirement is largely a response to a diminishing *product life time*, where consumers replace old products much more frequently due

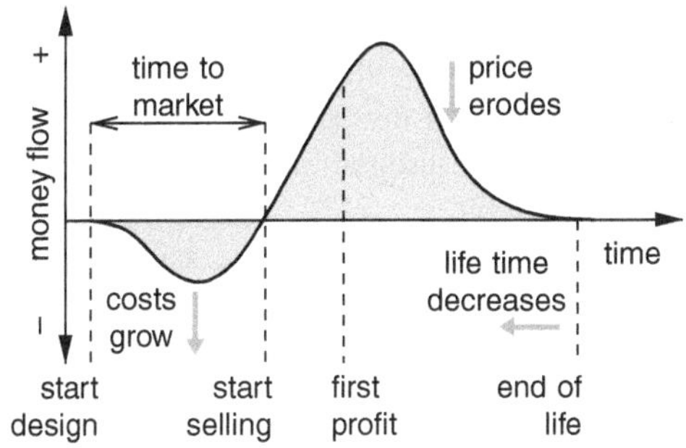

Fig. 1.3 Product life cycle

to rapid technology changes [89]. Mobile phone manufacturers, for example, release two major product lines per year compared with one just a few years ago [80]. Furthermore, as the product life time decreases, the units sold must still generate enough profit to cover the *rising costs* of manufacturing and design.

Profit is a major concern, as the manufacturing *Non-Recurring Engineering (NRE)* cost for a contemporary SoC is in the order of $1M, and design NRE cost routinely reaches $10M to $100M [112]. Traditionally, the rising costs are mitigated by high volumes and a high degree of system integration of heterogeneous technologies [88]. However, an increasing portion of the NRE cost is going into design and test and in 2007, for the first time in the history of SoC design, software design cost exceeded hardware design cost, and now accounts for 80% or more of the embedded systems development cost [89]. Thus, we see a steep increase in NRE cost for SoC designs which is not compensated for by higher volumes or higher margins. On the contrary, volumes are increasingly dependent on a low TTM, and margins decrease as prices erode quicker over time. The *International Technology Roadmap for Semiconductors (ITRS)* [89] goes as falr as saying that *the cost of designing the SoC is the greatest threat to continuation of the semiconductor road map*.

1.1.4 Summary

To summarise the trends, we see growing needs for functionality and performance, coupled with increasingly diverse requirements for different applications. Additionally, applications are becoming more dynamic, and are started and stopped at run time by the user. The diversity in functionality and requirements is also reflected in the architecture, as the applications are implemented by heterogeneous hardware and software IP, typically from multiple independent design teams. The increasing number of applications and resources also lead to an increased amount of resource sharing, both within and between use-cases. We illustrate all these trends in the following section, where we introduce an example system that we refer to throughout this work.

1.1.5 Example System

The system in Fig. 1.4 serves as our design example. Despite its limited size, this system is an example of a software-programmable, highly-parallel Multi-Processor SoC (MPSoC), as envisioned by, e.g., [80, 88, 112]. The system comprises a host processor, a number of heterogeneous processing engines, peripherals and memories. Three processors, a *Very Long Instruction Word (VLIW)*, an ARM and a μBlaze,[1] are connected to a memory-mapped video subsystem, an embedded SRAM, an audio codec and a peripheral tile with a character display, push buttons and a touch-screen controller. The host controls all the IPs and the *interconnect*, which binds them all together.

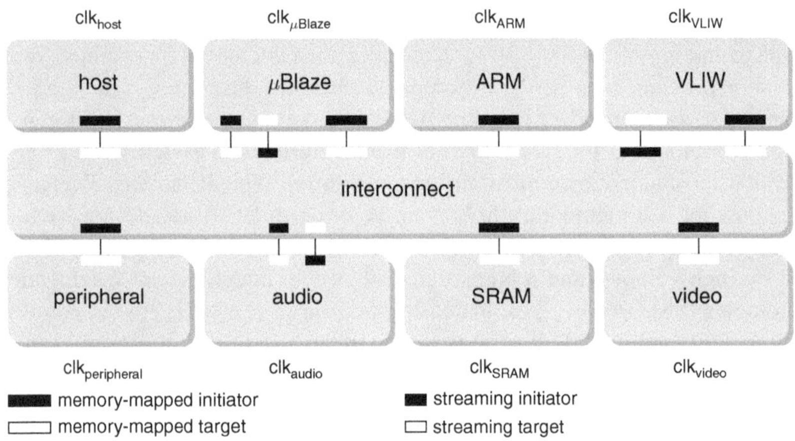

Fig. 1.4 Example system

For our example we assume a static set of applications, defined at design time, with tasks already statically mapped onto hardware IPs. This assumption is in line with the design flow proposed in this work. We consider six applications running on this system, as illustrated in Fig. 1.5. First, the two audio applications, both with firm real-time requirements, as failure to consume and produce samples at 48 kHz causes noticeable clicks and sound distortion. Second, a Motion-JPEG (M-JPEG) decoder and a video game. These applications both have soft real-time requirements, as they have a frame rate that is desirable, but not critical, for the user-perceived quality. Lastly, the two applications that run on the host. In contrast to the previous applications, they have no real-time constraints and only need to be functionally correct. In Fig. 1.5, the individual tasks of the applications are shown above and below the IPs to which they are mapped, and the communication between IPs is indicated by

[1] The names are merely used to distinguish the three processors. Their actual architecture is of no relevance for the example. The three processor families have, however, been demonstrated together with the interconnect proposed in this work.

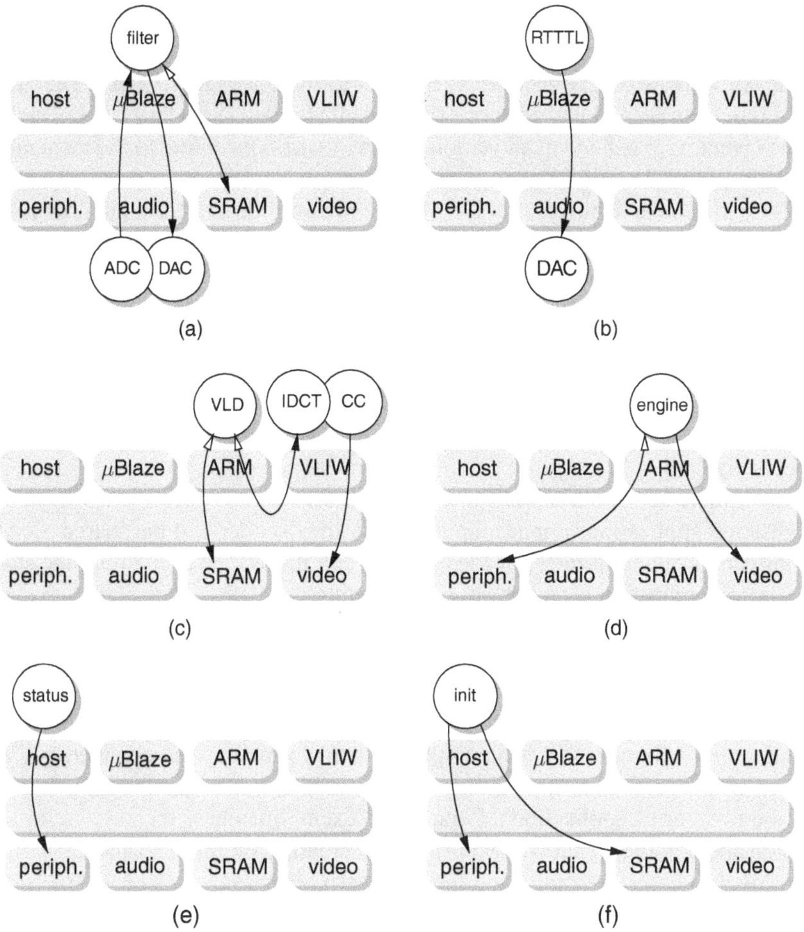

Fig. 1.5 Example applications and their mappings. (**a**) Audio filter application. (**b**) Ring-tone player application. (**c**) M-JPEG decoder application. (**d**) Video game application. (**e**) Status and control application. (**f**) Initialisation application

arrows. The solid-and open-headed arrows denote requests and responses, respectively. We now discuss the different applications in-depth.

The audio filter task runs on the μBlaze.[2] Input samples are read from the audio line-in Analog to Digital Converter (ADC) and a reverb effect is applied by adding attenuated past output samples that are read back from the SRAM. The output is written back to the line-out Digital to Analog Converter (DAC) and stored in the SRAM for future use. The Ring Tone Text Transfer Language (RTTTL) player,

[2] For brevity, we assume that the processors have local instruction memories and leave out the loading of these memories. In Chapter 7 we demonstrate how loading of instructions is taken into account.

running on the same μBlaze, interprets and generates the waveforms corresponding to ring-tone character strings and sends the output to the audio line-out.

The software M-JPEG decoder is mapped to the ARM and VLIW. The ARM reads input from SRAM and performs the Variable Length Decoding (VLD), including the inverse zig-zag and quantisation, before it writes its output to the local memory of the VLIW. The VLIW then carries out the Inverse Discrete Cosine Transform (IDCT) and Colour Conversion (CC) before writing the decoded pictures to the video output, eventually to be presented to the display controller. Both the ARM and the VLIW make use of *distributed shared memory* for all communication, and rely on *memory consistency models* to implement synchronisation primitives used in inter-processor communication. The other soft real-time application in our system is a video game. The ARM is responsible for rendering the on-screen contents and updating the screen based on user input. Memory-mapped buttons and timers in the peripheral tile are used to provide input and calibrate the update intervals, respectively.

Lastly, there are two non-real-time applications that both involve the host. The initialisation application supplies input data for the decoder into the shared SRAM, and initialises sampling rates for the audio. When the system is running, the status and control application is responsible for updating the character display with status information and setting the appropriate gain for the audio input and output.

The individual *applications* are combined into *use-cases*, based on constraints as shown in Fig. 1.6a. An edge between two applications means that they may run concurrently. From these constraints we get a set of use-cases, determined by the cliques (every subgraph that is a complete graph) formed by the applications. The number of use-cases thus depends on the constraints, and is typically much larger than the number of applications. Figure 1.6b exemplifies how the user could start and stop applications dynamically at run time, creating six different use-cases. Traditionally, use-cases are optimised independently. However, as seen in the figure, applications typically span multiple use-cases, e.g. the filter application continues to run as the decoder is stopped and the game started. Even during such changes, the real-time requirements of the filter application must not be violated.

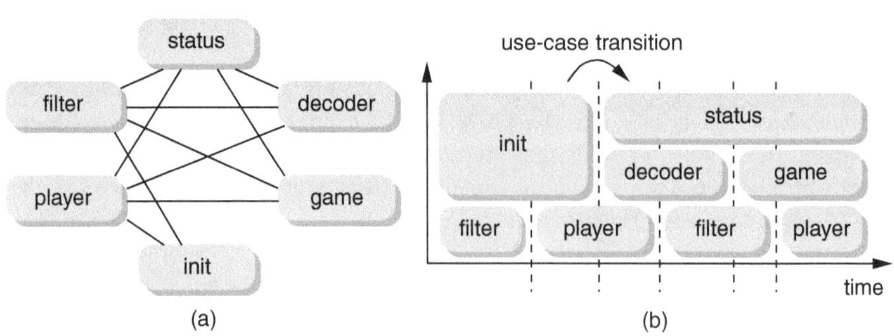

Fig. 1.6 Example use-cases. (**a**) Example use-case constraints. (**b**) Example use-case transitions

Already in this small system, we have multiple *concurrent applications*, and *a mix of firm-, soft- and non-real-time* requirements. The tasks of the applications are *distributed* across multiple heterogeneous resources (as in the case of the decoder application), and the resources in turn (in this case the interconnect, SRAM and peripheral tile) are *shared* between applications. The hardware IPs make use of both bi-directional address-based *memory-mapped protocols* such as DTL [49], AXI [8], OCP [147], AHB [2], PLB [160], Avalon-MM [4] (on the VLIW, ARM, μBlaze, video, SRAM and peripheral) and uni-directional *streaming protocols* such as DTL PPSD [49], Avalon-ST [4] and FSL [54] (on the μBlaze and audio). Additionally, the system has *multiple clock domains*, as every IP (and also the interconnect) resides in a clock domain of its own.

Having introduced and exemplified the trends in embedded systems, we continue by looking at the consequences of the system design and implementation.

1.2 Requirements

The trends have repercussions on all parts of the system, but are especially important for the design and implementation of the interconnect. The interconnect is a major contributor to the time required to reach timing closure on the system [89]. On the architectural level, the increasing number of IPs translates directly to a rapidly growing parallelism that has to be delivered by the interconnect in order for performance to scale. The interconnect is also the location where the diverse IP interfaces must interoperate. Concerning the increasing integration and verification complexity, the interconnect plays an important role, as it is the primary locus of the interactions between applications. The interconnect is also central in enabling real-time guarantees for applications where the tasks are distributed across multiples IPs. The impact the interconnect has makes it a key component of the MPSoC design [112, 162].

We cannot change the applications or the innovation speed, but instead can look at *simplifying the design and verification process* through the introduction of aelite, a new on-chip interconnect. We believe that the challenges introduced require a platform template that offers:

- *scalability* at the physical and architectural level (Section 1.2.1), allowing a large number of applications and IPs to be integrated on the SoC;
- *diversity* in IP interfaces and application communication paradigms (Section 1.2.2) to accommodate a variety of application behaviours implemented using heterogeneous IP from multiple vendors;
- *composability* of applications (Section 1.2.3), enabling independent design and verification of individual applications and applications as a unit of reuse;
- *predictability* with lower bounds on performance (Section 1.2.4), enabling formal analysis of the end-to-end application behaviour for real-time applications;
- *reconfigurability* of the hardware platform (Section 1.2.5), accommodating all use-cases by enabling dynamic starting and stopping of applications; and

- *automation* of platform mapping and synthesis (Section 1.2.6) to help the designer go from high-level requirements to an implementation.

We now explain each of the requirements in more detail before detailing how this work addresses them.

1.2.1 Scalability

The growing number of applications leads to a growing number and larger heterogeneity of IPs, introducing difficulties in timing validation and in connecting blocks running at different speeds [12]. This calls for *scalability at the physical level*, i.e. the ability to grow the chip size without negatively affecting the performance.

To enable components which are externally delay insensitive [35], Globally Asynchronous Locally Synchronous (GALS) design methods are used to decouple the clocks of the interconnect and the IPs [27, 62, 154], thus facilitating system integration [103]. Active rather than combinational interconnects are thus needed [201] to decouple computation and communication [96, 173], i.e. through the introduction of delay-tolerant protocols for the communication between IPs.

GALS at the level of IPs is, however, not enough. The interconnect typically spans the entire chip, and existing bus-based interconnects have many global wires and tight constraints on *clock skew*. With the increasing die sizes, it also becomes necessary to relax the requirements on synchronicity *within* the interconnect [154]. *Networks on Chip* (NoC) alleviate those requirements by moving from synchronous to mesochronous [29, 78, 154] or even asynchronous [12, 27, 165] communication. To achieve scalability at the physical level *we require GALS design at the level of independent IPs, and a mesochronous (or asynchronous) interconnect.*

Physical scalability is of no use unless the platform is *scalable at the architectural level*, i.e. supports a growing number of IPs and logical interconnections without negatively affecting the performance. Existing bus-based solutions address the problem by introducing more parallelism with outstanding transactions, and improvements such as bridges and crossbars [162]. NoCs extend on these concepts with their modular design, reuse, homogeneity and regularity [19, 43], offering high throughput and good power efficiency [12]. For architectural scalability *we require a modular interconnect without inherent bottlenecks.*

Scalability at the physical and architectural levels is necessary but not enough. The growing number of IPs and applications leads to a growing *diversity* in interfaces and programming models that the interconnect must accommodate.

1.2.2 Diversity

As we have seen, applications have diverse behaviours and requirements. Applications like the filter in our example system have firm real-time requirements, but also a fairly static behaviour. The M-JPEG player, on the other hand, is highly dynamic due to the input-dependent behaviour. This is also reflected in its more relaxed soft real-time requirements. To facilitate application diversity, *we require that applications are not forced to fit in a specific formal model*, e.g. have design-time schedules.

To facilitate the increased parallelism, diversity is also growing in the programming models and communication paradigms used by different IPs. SoCs are evolving in the direction of distributed-memory architectures, where the processor tiles have local memories, thus offering high throughput and low latency [140, 180], coupled with a low power consumption [141]. As the distributed nature of the interconnect leads to increased latencies, it is also important that the programming model is *latency tolerant*, i.e. enables maximal concurrency. In addition to memory-mapped communication, *streaming communication* (message passing) between IPs is growing in importance to alleviate contention for shared memories and is becoming a key aspect in achieving efficient parallel processing [112]. Hence, *we require that the interconnect offers both streaming (message passing) and distributed shared memory communication, and an established memory consistency model.*

The hardware IPs from different vendors use different *interfaces*, e.g. AHB [2] or AXI [8] for processors from ARM, PLB [160] for the μBlaze family from Xilinx, and DTL [49] for IP from Philips and NXP. To enable the use of existing IP, *we require that the interconnect supports one or more industry-standard interfaces, and that it is easy to extend to future interfaces.*

Lastly, there is also diversity in how the interconnect hardware and software is used by the system designer. That is, the tools used for, e.g., compilation, elaboration, simulation and synthesis differ between users, locations and target platforms. As a consequence, *we require that the interconnect uses standard hardware-description languages and programming languages for the interconnect hardware and software, respectively.*

Not only are applications diverse, they also share resources. Hence, for verification of their requirements, there is a need to decouple their behaviours as described next.

1.2.3 Composability

With more applications and more resources, the level of resource sharing is increasing. The increased sharing causes *more application interference* [143, 166], causing systems to behave in what has been described as *mysterious ways* [55]. In the presence of functional and non-functional application requirements the interference has severe implications on the effort involved in verifying the requirements, and increasing dynamism within and between applications exacerbates the problem [152]. Although individual IPs are pre-verified, the verification is usually done in a limited context, with assumptions that may not hold after integration. As a result, the complexity of *system* verification and integration grows exponentially with the number of applications, far outgrowing the incremental productivity improvement of IP design and verification [166], and making system-level simulation untenable [152]. Higher levels of abstraction, e.g. transaction-level modelling, mitigate the problem but may cause bugs to disappear or give inaccurate performance measures. The high-level models are thus not able to guarantee that the application requirements are met. Instead, we need to develop the technology that allows applications to be easily *composed* into working systems, independent of other applications in the system [9, 77, 90, 146].

Composability is a well-established concept in systems used in the automotive and aerospace domains [6, 169]. In a composable platform one application cannot change the behaviour of another application.[3] Since application interference is eliminated, the resources available before and after integration can only be different due to the intrinsic *platform uncertainty*, caused by, e.g., clock domain crossings. Composability enables design and debugging of applications to be done in isolation, with higher simulation speed and reduced debugging scope [77]. This is possible as only the resources assigned to the application in question have to be included in the simulation. Everything that is not part of the application can be safely excluded. As a result, probabilistic analysis, e.g. average-case performance or deadline miss rate, during the application design gives a good indication of the performance that is to be expected after integration. Application composability also improves IP protection, as the functional and temporal behaviour before and after integration is independent of other applications. Consequently, there is never a reason blame other applications for problems, e.g. violated requirements in the form of bugs of deadline misses. As a result, the IP of different *Independent Software Vendors* (ISV) do not have to be shared, nor the input stimuli.

Application composability places stringent requirements on the management of all resources that are shared by multiple applications, in our case the interconnect.[4] Composable sharing of resources requires *admission control* coupled with *non-work-conserving arbitration* [205] between applications, where the amount of resources and time resources are available, is not influenced by other applications. For a shared memory controller, for example, the exact number of cycles required to finish a request must depend only on the platform and the cycle in which the request is made (and previous requests from the same application), and not the behaviour of other applications [109].

With application composability, the quality of service, e.g. the deadline miss rate, and bugs, e.g. due to races, are unaffected by other applications. We already expect this type of composable behaviour on the level of individual transistors, gates, arithmetic logic units and processors [166]. Taking it one step further, in this work *we require that applications can be viewed as design artefacts, implemented and verified independently, and composed into systems.*

---

[3] We return to discuss other aspects of composability in Chapter 9, when reviewing related work.

[4] Sharing of *processors* between applications is outside the scope of this work, a reasonable limitation that is elaborated on in Chapter 7 when discussing our example system.

Composability addresses the productivity and verification challenge by a *divide-and-conquer* strategy. It does, however, not offer any help in the verification of the real-time requirements of the individual applications. For this purpose, we need temporal *predictability* [14].

1.2.4 Predictability

We refer to an architectural component as *predictable* when it is able to guarantee (useful) *lower bounds* on performance, i.e. minimum throughput and maximum latency [63, 90], given to an application. Predictability is needed to be able to guarantee that real-time requirements are met for firm real-time applications, thus delivering, for example, a desired user-perceived quality or living up to the latency requirements in wireless standards.

Note that predictability and composability are *orthogonal* properties [77]. For an illustrative example, consider our example system, and then in particular the filter and decoder applications. If the interconnect and SRAM use composable arbitration, as later described in Chapter 3, but the μBlaze and ARM are running a non-real-time operating system, then the platform is *composable and not predictable*, because the applications do not influence each other, but the operating system makes it difficult, if not impossible, to derive *useful bounds* on the worst-case behaviour. In contrast, if the μBlaze and ARM are used without any caches and operating systems, but the SRAM is shared round robin, then the platform is *predictable and not composable*. This is due to the fact that the applications influence each other in the shared resource, but lower bounds on the provided service can be computed.

Predictable resources are, however, not enough. To determine bounds on the temporal behaviour of an application, the architecture and resource allocations must be modelled conservatively according to a specific *Model of Computation* (MoC). Moreover, also the application (or rather its tasks) must fit in a MoC that allows analytical reasoning about relevant metrics. Using predictability for application-level analysis thus *places limitations* on the application. In addition, the MoC must be *monotonic* [161], i.e. a reduced task execution time cannot result in a reduction of the throughput or increased latency of the application [161]. Without monotonicity, a *decrease* of the execution time at the task level may cause an *increase* at the application level [64]. To benefit from the bounds provided by a predictable platform, the MoC (but not the implementation itself) must be free of these types of scheduling anomalies. For application-level predictability *we require that the interconnect guarantees a minimum throughput bound and maximum latency bound and that the temporal behaviour of the interconnect can be conservatively captured in a monotonic MoC.*

To enable composability and predictability in the presence of multiple use-cases, the allocations of resources to applications must be *reconfigurable* as applications are started and stopped.

1.2.5 Reconfigurability

As already illustrated in Fig. 1.6, applications are started and stopped at run time, creating many different use-cases. Reconfigurability is thus required to allow dynamic behaviour between applications, i.e. to modify the set of running applications as illustrated in Fig. 1.7. Moreover, applications must be started and stopped independently of one another [72, 101] to maintain composability (and predictability if implemented by the application in question) during reconfiguration. Consequently, *we require that one application can be started and stopped without affecting the other applications of the system.*

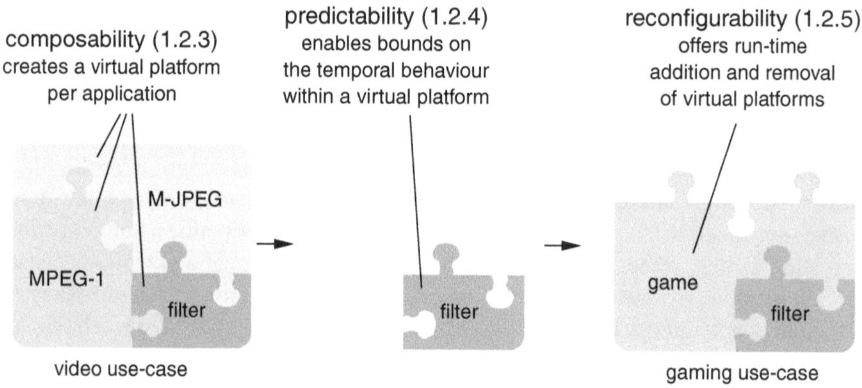

Fig. 1.7 Reconfigurable composability and predictability

The problem of reconfigurability, however, does not entail only the allocation of resources. The allocations must also be instantiated at run time in a safe manner. Internally, programming the interconnect involves modifications of the registers in individual components [28, 46]. To mitigate the complexity of reconfiguration, *the abstraction level must be raised* and provide an interface between the platform and the applications [111, 115]. Moreover, it is crucial that the changes are applied in such a way as to leave the system in a *consistent state*, that is, a state from which the system can continue processing normally rather than progressing towards an error state [102]. Simply updating the interconnect registers could cause out-of-order delivery or even no delivery, with an erroneous behaviour, e.g. deadlock as the outcome. In addition to correctness, some application transitions require *an upper bound* on the time needed for reconfiguration. Consider for example the filter and ring-tone player in our example system where a maximum period of silence is allowed in the transition between the two applications. To accommodate these applications, *we require a run-time library that hides the details of the architecture and provides correct and timely reconfiguration of the interconnect.*

As we have seen, a scalable, diverse, composable, predictable and reconfigurable on-chip interconnect addresses many problems, but also pushes decisions to design time. However, the design effort cannot be increased, and must remain at current levels for the foreseeable future [88]. Already today, meeting schedules is the number one concern for embedded developers [189]. As a consequence, the degree of *automation*, particularly in verification and implementation, must be increased.

1.2.6 Automation

Automation is central for the *functional scalability* of the interconnect, i.e. the ability to scale with increasing requirements, and is primarily needed in three areas. First, in platform *dimensioning* and resource *allocation*. That is, going from high-level requirements and constraints, formulated by the application designers to a specification of interconnect architecture and allocations. Second, in *instantiation* of the interconnect hardware and software, e.g. going from the specification to a functional system realisation in the form of SystemC or RTL HDL. Third, in the *evaluation* of the system by automating performance analysis and cost assessment. Throughout these steps, it is important that the automated flow allows user interaction at any point.

Deferring decisions to run time typically enables higher performance,[5] but also makes real-time analysis more difficult, if not impossible [109]. Compare, for example, the compile-time scheduling of a VLIW with that of a super-scalar out-of-order processor. Similar to the VLIW, the predictability and composability in our platform is based on hard resource reservations made at design and compile time. Therefore, platform mapping, essential for all SoCs [89], is even more prominent in our system. As part of the interconnect design flow, *we require a clear unambiguous way to express the application requirements*. Furthermore, *we require that the design flow is able to automatically turn the requirements into an interconnect specification* by performing the resource dimensioning and resource allocation.

Once we have a specification of the hardware and software that together constitute the platform, it must be instantiated. This involves turning the hardware specification into executable models or industry-standard hardware description languages, to be used by, e.g., Field-Programmable Gate Array (FPGA) or Application-Specific Integrated Circuit (ASIC) synthesis tools. Additionally, the hardware must be accompanied by the software libraries required to program and control the hardware, and a translation of the allocation into a format useful to these libraries. Furthermore, this software must be portable to a wide range of host implementations, e.g. different processors and compilers, and be available for simulation as well as actual platform instances. Thus, *we require that the design flow instantiates a complete hardware and software platform, using industry-standard languages*.

Lastly, automation is required to help in the evaluation of the interconnect instance. The design flow must make it possible to evaluate application-level performance, either through simulation or formal analysis, and do so at multiple levels,

[5] More (accurate) information is available, but with smaller scope (local). Run-time processing is also usually (much) more constrained than design-time processing in terms of compute power.

ranging from transaction-level to gate-level. It is also essential that the tools enable the designer to assess the cost of the system, for example, by providing early area estimates. To enable the efficient design, use and verification of tomorrow's billion-transistor chips, *we require that the interconnect offers both (conservative) formal models and simulation models, and that the latter enable a trade-off between speed and accuracy*.

1.3 Key Components

To address the aforementioned requirements, in this work, we:

- Identify the *requirements for composability and predictability* in multi-application SoCs [77] (this chapter).
- Propose a complete on-chip *interconnect architecture* [69] with local buses, protocol shells, clock bridges, network interfaces (NI), routers and links [78], together with a control infrastructure for run-time reconfiguration [68] (Chapter 3).
- Provide a *resource allocation algorithm* for composability and predictability [70, 74] across multiple use-cases [72], coupled with sharing of time slots [71] within a use-case (Chapter 4).
- Deliver *run-time libraries for the run-time instantiation* of the resource allocations [68], ensuring correctness and giving temporal bounds on reconfiguration operations (Chapter 5).
- Present *formal models of a network channel*, and show how to use them for buffer sizing [75], and for application-level performance guarantees [79] (Chapter 6).
- Demonstrate the applicability of this work by constructing *an example multi-processor system instance*, with a diverse set of applications and computational cores, and map it to an FPGA [77] (Chapter 7).
- Evaluate the proposed interconnect and design flow using two large-scale multi-core consumer applications (Chapter 8)

The final part of the proposed aelite interconnect is the design flow depicted in Fig. 1.8. As shown in the figure, this work takes as its starting point the specification of the physical interfaces that are used by the IPs. The aforementioned architecture description is complemented by the communication requirements of the applications, specified per application as a set of logical interconnections between IP ports, each with bounds on the maximum latency and minimum throughput. The architecture and communication specification also allows the user of the design flow to constrain how the physical IPs are placed and how the applications are combined temporally. These specifications are assumed to be given by the application and system designer, referred to as the *user* in Fig. 1.8. The interface specifications follow directly from the choice of IPs, and the use-case constraints are part of the early system design. It is less obvious, however, where the requirements on the interconnect stem from. Depending on the application, the desired throughput and latency

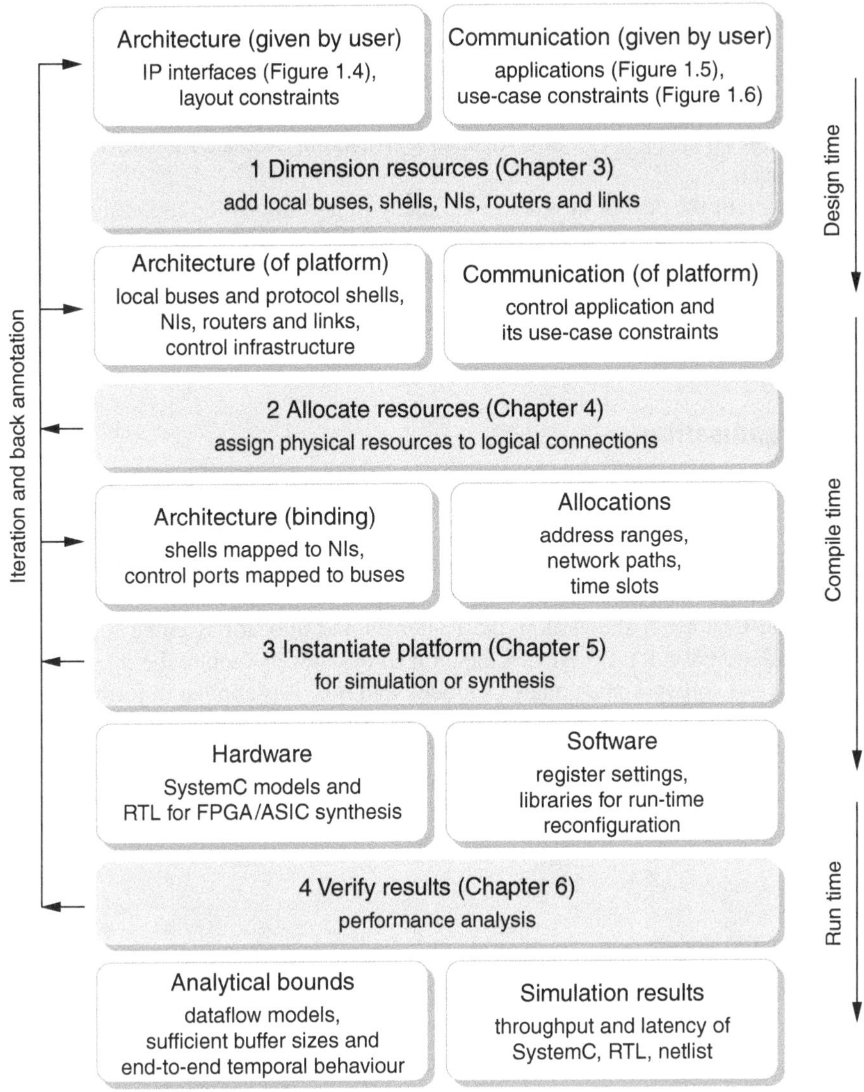

Fig. 1.8 Design-flow overview

may be analytically computed, the outcome of high-level simulation models of the application, or simply guesstimates based on back-of-the-envelope calculations or earlier designs. Automation of this step is outside the scope of this work, but we refer to related work that address the issue, and present a number of design examples that illustrate possible approaches.

The outcome of the proposed design flow is a complete SoC interconnect architecture and a set of resource allocations that together implement the requirements specified by the user. The resulting platform is instantiated (in synthesisable HDL or

SystemC), together with the software libraries required to configure and orchestrate use-case switches (in Tcl or ANSI C). Verification of the result is performed per application, either by means of simulation or by using conservative models of the interconnect. At this point, it is also important to note that the actual applications, whether run on an FPGA or analytically modelled are evaluated together with the interconnect.

Throughout the design flow, there are many opportunities for successive refinement, as indicated by the iteration in Fig. 1.8. Iteration is either a result of failure to deliver on requirements or a result of cost assessment, e.g. silicon area or power consumption. It is also possible that the specification of requirements changes as a result of the final evaluation.

1.4 Organisation

The remainder of this work is organised as follows. We start by introducing the key concepts of aelite in Chapter 2. Thereafter, Chapter 3 begins our bottom-up description of the solution, introducing the design-time dimensioning of the interconnect. This is followed by a discussion on the compile-time allocation of resources in Chapter 4. Chapter 5 shows how the resources and allocations come together in an instantiation at run time. Next, Chapter 6 shows how to capture the interconnect hardware and software in an analysis model, and how it is applied in formal verification of the end-to-end application behaviour. All the ingredients come together in Chapter 7, where a complete example multi-processor system is designed, instantiated and verified. In Chapter 8 we evaluate the entire interconnect design flow using two large industrial case studies. Finally, we review related work in Chapter 9, and end with conclusions and directions for future work in Chapter 10.

Chapter 2
Proposed Solution

In this chapter we give a high-level view of the building blocks of the interconnect and discuss the rationale behind their partitioning and functionalities. We start by introducing the blocks by exemplifying their use (Section 2.1). This is followed by a discussion of how the interconnect delivers scalability at the physical and architectural level (Section 2.2). Next, we introduce the protocol stack and show how it enables diversity, both in the application programming models and in the IP interfaces (Section 2.3). We continue by explaining how we provide temporal composability when applications share resources (Section 2.4). Thereafter, we detail how our interconnect enables predictability on the application level by providing dataflow models of individual connections (Section 2.5). Next we show how we implement reconfigurability to enable applications to be independently started and stopped at run time (Section 2.6). Lastly, we describe the rationale behind the central role of automation in our proposed interconnect (Section 2.7), and end this chapter with conclusions (Section 2.8).

2.1 Architecture Overview

The blocks of the interconnect are all shown in Fig. 2.1, which illustrates the same system as Fig. 1.4, but now with an expanded view of the interconnect, dimensioned for the applications in Fig. 1.5 and use-cases in Fig. 1.6.

To illustrate the functions of the different blocks, consider a load instruction that is executed on the ARM. The instruction causes a bus *transaction*, in this case a read transaction, to be initiated on the data port of the processor, i.e. the *memory-mapped initiator* port. Since the ARM uses distributed memory, a *target bus* with a *reconfigurable address decoder* forwards the read *request message* to the appropriate initiator port of the bus, *based on the address*. The *elements* that constitute the request message, e.g. the address and command flags in the case of a read request, are then serialised by a *target shell* into individual words of streaming data, as elaborated on in Section 2.3. The streaming data is fed via a *Clock Domain Crossing* (CDC) into the NI *input queue* corresponding to a specific *connection*. The data items reside in that input queue until the *reconfigurable NI arbiter* schedules

A. Hansson, K. Goossens, *On-Chip Interconnect with Aelite*, Embedded Systems,
DOI 10.1007/978-1-4419-6865-4_2,

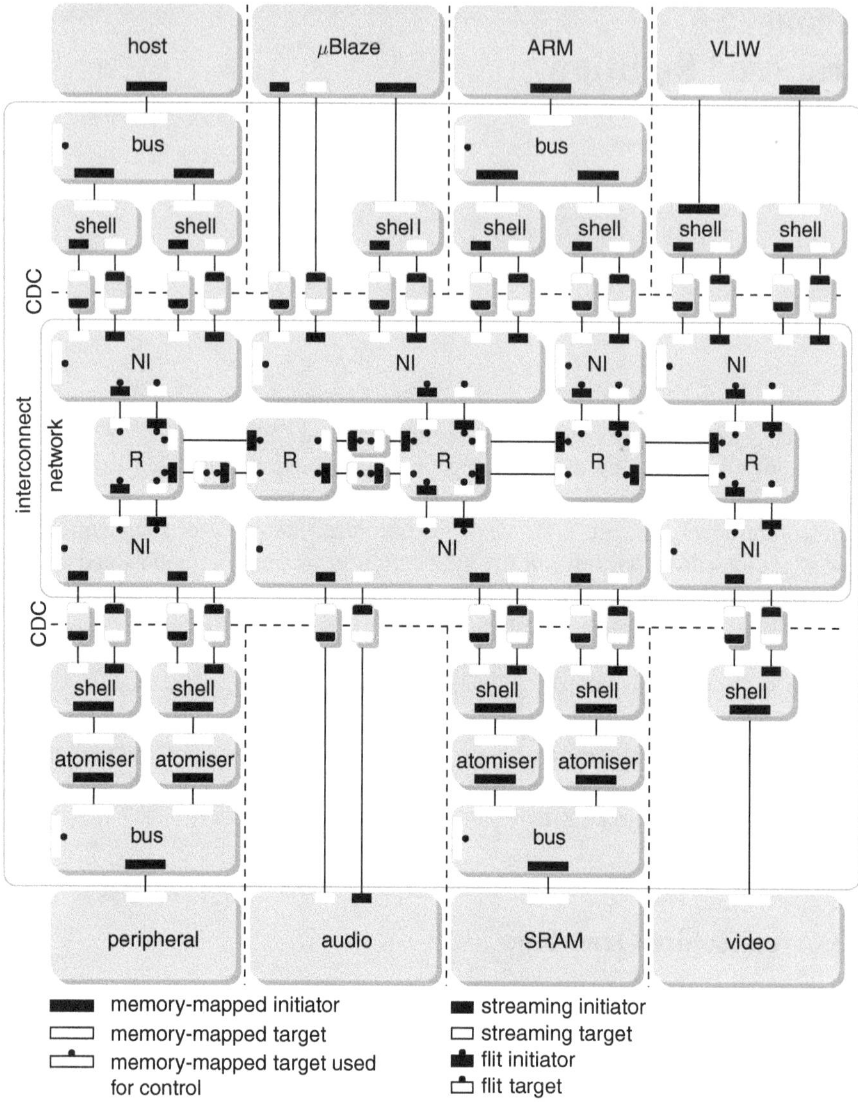

Fig. 2.1 Interconnect architecture overview

the connection. The streaming data is *packetised* and injected into the router network, as *flow-control digits (flits)*. Based on a *path* in the *packet header*, the flits are forwarded through the router network, possibly also encountering *pipeline stages* on the links, until they reach their destination NI. In the network, the flits are forwarded *without arbitration*, as discussed further in Section 2.1.1. Once the flits reach the destination NI, their payload, i.e. the streaming data, is put in the NI *output queue* of the connection and passes through a clock domain crossing into an *initiator shell*.

The shell represents the ARM as a memory-mapped initiator by reassembling the request message. If the destination target port is not shared by multiple initiators, the shell is directly connected to it, e.g. the video tile in Fig. 2.1. For a shared target, the request message passes through an *atomiser*. By splitting the request, the atomiser ensures that transactions, from the perspective of the shared target, are of a *fixed size*. As further discussed in Section 2.4, the atomiser also provides buffering to ensure that atomised transaction can complete in their entirety *without blocking*. Each atomised request message is then forwarded to an *initiator bus* that *arbitrates* between different initiator ports. Once granted, the request message is forwarded to the target, in this case the SRAM, and a *response message* is generated. The elements that constitute the response message are sent back through the bus. Each atomised response message is buffered in the atomiser that reconstructs the entire response message. After the atomiser, the response message is presented to the initiator shell that issued the corresponding request. The shell adds a message header and serialises the response message into streaming data that is sent back through the *network*, hereafter denoting the NIs, routers and link pipeline stages. On the other side of the network, the response message is reassembled by the target shell and forwarded to the bus. The target bus implements the transaction *ordering* corresponding to the IP port protocol. Depending on the protocol, the response message may have to wait until transactions issued by the ARM before this one finish. The bus then forwards the response message to the ARM, completing the read transaction and the load instruction. In addition to response ordering, the target bus also implements mechanisms such as tagging [49] to enable programmers to choose a specific *memory-consistency model*.

In the proposed interconnect architecture, the applications share the network and possibly also the memory-mapped targets.[1] Arbitration thus takes place in the NIs and the initiator buses. Inside the network, however, *contention-free routing* [164] removes the need for any additional arbitration. We now discuss the concepts of contention-free routing in more detail.

2.1.1 Contention-Free Routing

Arbitration in the network is done at the level of flits. The injection of flits is regulated by TDM *slot tables* in the NIs such that no two flits ever arrive at the same link at the same time. Network *contention and congestion* is thus avoided. This is illustrated in Fig. 2.2, where a small part of the network from our example system is shown. In the figure, we see three *channels*, denoted c_0, c_1 and c_2, respectively. These three channels correspond to the connections from the streaming port on the μBlaze to the DAC, the request channel from the memory-mapped port on the μBlaze to the SRAM, and the request channel from the ARM to the SRAM.

[1] Memory-mapped initiator ports and target buses are not shared in the current instance, and is something we consider future work.

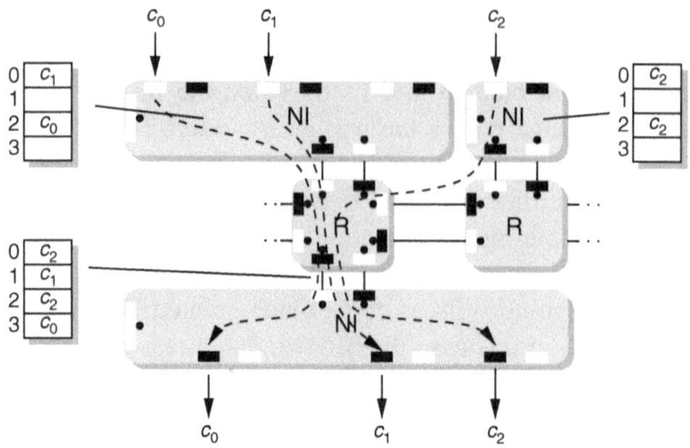

Fig. 2.2 Contention-free routing

Channels c_0 and c_1 have the same source NI and have slots 2 and 0 reserved, respectively. Channel c_2 originates from a different NI and also has slots 0 and 2 reserved. The TDM table size is the same throughout the network, in this case 4, and every slot corresponds to a flit of fixed size, assumed to be three words throughout this work.

For every hop along the path, the reservation is shifted by one slot, also denoted a *flit cycle*. For example, in Fig. 2.2, on the last link before the destination NI, c_0 uses slot 3 (one hop) c_1 slot 1 (one hop) and c_2 slots 0 and 2 (two hops). The notion of a flit cycle is a result of the alignment between the *scheduling interval* of the NI, the flit size and the forwarding delay of a router and link pipeline stage. As we shall see in Chapter 4, this alignment is crucial for the resource allocation.

With contention-free routing, the router network behaves as a non-blocking pipelined multi-stage switch, with a global schedule implied by all the slot tables. Arbitration takes place once, at the source NI, and not at the routers. As a result each channel acts like an independent FIFO, thus offering composability. Moreover, contention-free routing enables predictability by giving per-channel bounds on latency and throughput. We return to discuss composability and predictability in Sections 2.4 and 2.5, respectively. We now discuss the consequences of the scalability requirement.

2.2 Scalability

Scalability is required both at the physical and architectural level, i.e. the interconnect must enable both large die sizes and a large number of IPs. We divide the discussion of our proposed interconnect accordingly, and start by looking at the physical level.

2.2.1 Physical Scalability

To enable the IPs to run on independent clocks, i.e. GALS at the level of IPs, the NIs interface with the shells (IPs) through clock domain crossings. We choose to implement the clock domain crossings using bi-synchronous FIFOs. This offers a clearly defined, standardised interface and a simple protocol between synchronous modules, as suggested in [103, 116]. Furthermore, a clock domain crossing based on bi-synchronous FIFOs is robust with regards to metastability, and allows each locally synchronous module's frequency and voltage to be set independently [103]. The rationale behind the placement of the clock domain crossings between the NIs and shells is that all complications involved in bridging between clock domains are confined to a single component, namely the bi-synchronous FIFO. Even though the IPs have their own clock domains (and possibly voltage islands), normal test approaches are applicable as the test equipment can access independent scan chains per synchronous block of the system [103], potentially reusing the functional interconnect as a test access mechanism [20].

In addition to individual clock domains of the IPs, the link pipeline stages enable *mesochronous clocking* inside the network [78]. The entire network thus uses the same clock (frequency), but a phase difference is allowed between neighbouring routers and NIs. Thus, in contrast to a synchronous network, restrictions on the phase differences are relaxed, easing the module placement and the clock distribution. As the constraints on the maximum phase difference are only between neighbours, the clock distribution scales with the network size, as demonstrated in [83]. The rationale behind choosing a mesochronous rather than an asynchronous interconnect implementation is that it can be conceived as globally synchronous on the outside [29]. Thereby, the system designer does not need to consider its mesochronous nature. Furthermore, besides the clock domain crossings and link pipeline stages, all other (sub-)components of the interconnect are synchronous and consequently designed, implemented and tested independently.

2.2.2 Architectural Scalability

The physical scalability of the interconnect is necessary but not sufficient. The distributed nature of the interconnect enables architectural scalability by *avoiding central bottlenecks* [19, 43]. More applications and IPs are easily added by expanding the interconnect with more links, routers, NIs, shells and buses. Furthermore, the latency and throughput that can be offered to different applications is a direct result of the amount of contention in the interconnect. In the network, the *contention can be made arbitrarily low*, and links (router ports), NIs and routers can be added up to the level where each connection has its own resources, as exemplified in Fig. 2.3. In Fig. 2.3a, we start with two connections sharing an NI. We then add two links (router ports) in Fig. 2.3b (to an already existing router), thus reducing the level of contention. Figure 2.3c continues by giving each connection a dedicated NI. Lastly,

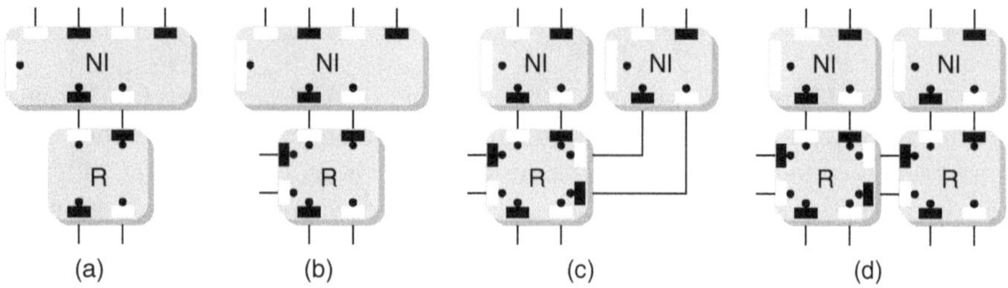

Fig. 2.3 Architecture scaling by adding links (**b**), NIs (**c**), and routers (**d**)

Fig. 2.3d distributes the connections across multiple routers, thus reducing the contention to a minimum (but increasing the cost of the interconnect). As we shall see in Section 2.4 the proposed interconnect also contributes to architectural scalability by having a router network that does not grow with the number of connections.

Outside the network, the contention for a shared target port, such as an off-chip memory controller, cannot be mitigated by the interconnect as it is inherent in the applications. This problem has to be addressed at the level of the programming model, which we discuss in depth in the following section. We return to discuss scalability in Chapter 8 where we demonstrate the *functional scalability* using two large-scale case studies, and also see concrete examples of the effects of inherent sharing.

2.3 Diversity

Diversity in interfaces and programming models is addressed by the protocol stack shown in Fig. 2.4. The stack is divided into five layers according to the seven-layer Open Systems Interconnection (OSI) reference model [45]. As seen in the figure, the memory-mapped protocol, the streaming protocol and the network protocol each have their own stack, bridged by the shell and the NI. We discuss the three stacks in turn.

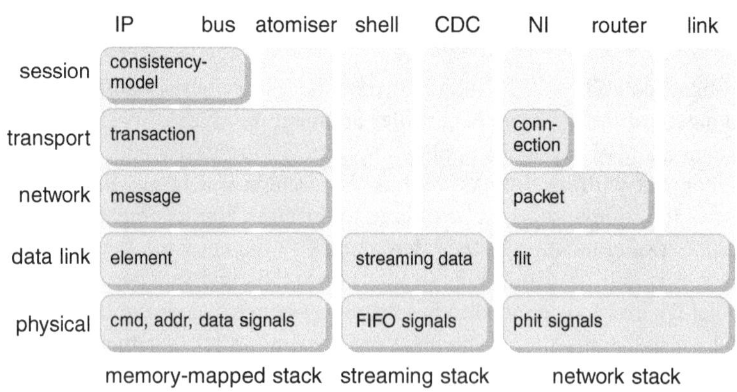

Fig. 2.4 Interconnect protocol stack

2.3.1 Network Stack

The network stack is similar to what is proposed in [18, 25, 118, 123]. The NI is on the transport level as it maintains end-to-end (from the perspective of the network stack) *connections* and guarantees ordering within, but not between connections [167]. A connection is a bi-directional point-to-point inter-connection, between two pairs of initiator and target streaming ports on the NIs. Two uni-directional *channels*, one in each direction, connect the two pairs of ports. Due to the bi-directional nature of a connection, streaming ports always appear in pairs, with one initiator and one target port. As we shall see in Section 2.4, the way in which connections are buffered and arbitrated is central to the ability to provide composable and predictable services in the network, and has a major impact on the NI and router architecture. As seen in Fig. 2.4, the router is at the network level and performs switching of *packets*. The last element of the network, the link pipeline stage, is at the data link level and is responsible for the (synchronous or mesochronous) clock synchronisation and flow control involved in the transport of *flits*. The physical layer, i.e. timing, bus width and pulse shape, is governed by the *phit* (physical digit) format.

From the perspective of the IPs, the network behaves as a collection of distributed and independent FIFOs (or virtual wires), with data entering at a streaming target port, and at a later point appearing at a corresponding streaming initiator port (determined by the allocation). Thus, the network stack is completely hidden from the IPs, and only used by means of the streaming stack.

2.3.2 Streaming Stack

The streaming stack is far simpler than the network stack, and only covers the two lowest layers. The NI, clock domain crossing, shell and IPs with streaming ports (like the μBlaze in our example system or the video blocks in [183]) all make direct use of this stack. The data-link level governs the flow control of individual words of *streaming data*. The streaming ports make use of a simple FIFO interface with a valid and accept handshake of the data. The physical level concerns the FIFO signal interface. For robustness, the streaming interfaces use *blocking flow control* by means of back pressure. That is, writing to a streaming target port that is not ready to accept data (e.g. due to a full FIFO) or reading from a streaming target port that has no valid data (e.g. due to an empty FIFO) causes a process to stall. We return to discuss the blocking flow control and its implications in Section 2.4.

As illustrated in Fig. 2.5, the NI bridges between the streaming stack and network stack by establishing connections between streaming ports and embedding streaming data in network packets. The connections are thus, via the streaming stack, offering bi-directional point-to-point streaming communication, *without any interpretation or assumptions* on the time or value of the individual words of streaming data. Moreover, with FIFO ordering inside a network channel, it is up to the users of the network, e.g. IPs and buses to implement a specific inter-channel ordering.

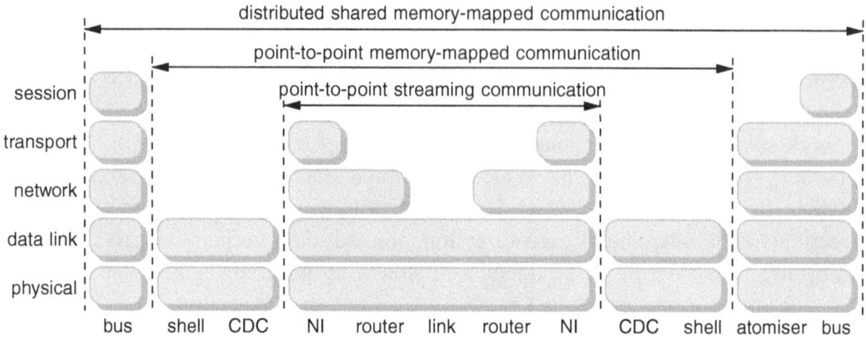

Fig. 2.5 Bridging the protocol stacks

Both the aforementioned properties are key in enabling diversity in communication paradigms and programming models, as described below.

The most basic use of the streaming stack is exemplified in Fig. 2.1 by the μBlaze that communicates with the audio directly via the streaming ports of the NI. One connection is sufficient to interconnect the two blocks, with one channel carrying data from the ADC to the μBlaze and the other channel from the μBlaze to the DAC. Only raw data, i.e. signed pulse-code-modulated samples, is communicated across both channels that are completely homogeneous [77].

2.3.3 Memory-Mapped Stack

In contrast to the simple FIFO interface of the streaming ports, memory-mapped protocols are based on a request–response transaction model and typically have interfaces with dedicated groups of wires for command, address, write data and read data [2, 4, 8, 49, 147, 160]. Many protocols also support features like byte enables and burst transactions (single request multiple data elements). The block that bridges between the memory-mapped ports of the IPs and the streaming ports of the NIs is a *shell* that (independently) serialises the memory-mapped request and response messages. As illustrated in Fig. 2.5, the shells enable *point-to-point memory-mapped communication* by bridging between the *elements* of a bus-protocol message, e.g. the address, command flags or individual words of write data, and words of streaming data by implementing the data-link protocol of both the bus stack and the streaming stack. As already mentioned, the network places no assumptions on the time and value of individual words of streaming data, and also not on the size, syntax, semantics or synchronisation granularity of messages. Consequently, the shells enable streaming communication and memory-mapped communication to co-exist in the interconnect. Moreover, different pairs of memory-mapped initiators and targets may communicate using different protocols, thus enabling multiple memory-mapped protocols.

The shells work on the granularity of a single memory-mapped initiator or target port. However, a memory-mapped initiator port often uses distributed memory [168], and accesses multiple targets, based on e.g. the address, the type of transaction (e.g. read or write) or dedicated identifier signals in the port interface [147]. The use of distributed memory is demonstrated by the host and ARM in Fig. 2.1. As the ARM accesses more than one target port, the outgoing requests have to be directed to the appropriate target, and the incoming responses *ordered* and presented to the initiator according to the protocol. In addition to the use of distributed memory at the initiator ports, memory-mapped target ports are often shared by multiple initiators, as illustrated by the SRAM in Fig. 2.1. A shared target must hence be *arbitrated*, and the initiators' transactions multiplexed according to the protocol of the target port. Next, we show how distributed and shared memory-mapped communication, as shown in Fig. 2.5, is enabled by the target buses and initiator buses, respectively.

Despite all the multiplexing and arbitration *inside the network* we choose to add buses *outside the network*. The primary reason why the network is not responsible for all multiplexing and arbitration is *the separation of bus and network stacks*. The network stack offers logical (bi-directional) point-to-point connections, without any ordering between connections. Ordering of, and hence synchronisation between, transactions destined for different memories depends on particular IP port protocols and is therefore addressed outside the network, in the target buses. The ordering and synchronisation between transactions place the buses at the session level in the stack, and adds a fifth layer to the interconnect protocol stack, as seen in Fig. 2.4. At the network level in the bus stack we have *messages* in the form of requests and responses. It is the responsibility of the bus to perform the necessary multiplexing and direct messages to the appropriate destination. Each message is in turn constructed of *elements* and the data-link layer is responsible for the flow control and synchronisation of such elements. Finally, the physical layer governs the different *signal groups* of the bus interface.

The division of the protocol stack clearly separates the network from any session-level issues and pushes those responsibilities to the buses. The network is thereby independent of the IP protocols and relies on target buses for distributed memory communication, and initiator buses for shared memory communication. The division enables the interconnect to support different types of bus protocols and different memory consistency models. The choice of a consistency model, e.g. *release consistency* [59], is left for the IP (and bus) developer. That is, the interconnect provides the necessary mechanisms, and it is left for the IPs to implement a specific policy. There are no restrictions on the consistency model, but to exploit the parallelism in the interconnect, it is important that the ordering is as relaxed as possible. The importance of parallelism, e.g. through transaction pipelining, is emphasised by the large (10–100 cycles best-case) latency involved in receiving response from a remotely located target.[2]

Note that the stack separation does not exclude the option to implement the protocol-related ordering and arbitration in the protocol shell, i.e. merge the bus

[2] If the ARM and SRAM in Fig. 2.1 run at 200 MHz and the network at 500 MHz, the best-case round-trip for a read operation is in the order of 30 processor cycles.

and shells as suggested in [168]. An integrated implementation may result in a more efficient design [91]. Indeed, placing both the shell and bus functionality inside such a block enables a design where the lower layers involved in the protocol translation functionality are shared by the ports, thus potentially achieving a lower hardware cost and latency. The main drawback with such a monolithic approach is the complexity involved in handling multiple concurrent transactions, while complying with both the bus protocol (on the session level) and streaming protocol. Our proposed division of the stack enables us to bridge between protocols on the lower layers, something that involves far fewer challenges than attempting to do so on the session level [118]. Furthermore, by having buses that complies with the IP protocol it is possible to reuse available library components. It is also possible to verify the buses independent of the network with established protocol-checking tools such as Specman [34], and to use any existing functional blocks for, e.g. word-width and endianness conversion, or instrumentation and debugging [192]. Lastly, thanks to the clear interface between shell and NI, the shells belonging to one IP port, e.g. the two shells of the ARM in Fig. 2.1, can easily be distributed over multiple NIs, e.g. to increase the throughput or provide lower latency.

2.4 Composability

As discussed in Chapter 1, we refer to a system where applications do not influence each other temporally as composable. Application composability is easily achieved by using dedicated resources for the different applications. This, however, is often too costly (in terms of power and area) or even impossible. Consider, for example, the case of an off-chip memory where pin constraints limit the number of memories. To fulfil the requirement on application composability, it is thus necessary to offer composable resource sharing by removing all interferences between applications. Four properties are central to the ability of providing composable sharing:

- the *flow-control scheme* used by the shared resource,
- the respective *granularities of flow control and arbitration*,
- the *size of an arbitration unit*,
- the *temporal interference* between arbitration units.

We now look at these properties in turn, using the abstract shared resource in Fig. 2.6 to illustrate the concepts. This resource corresponds to, e.g., the router network or the SRAM in Fig. 2.1.

2.4.1 Resource Flow-Control Scheme

Flow control is the process of adjusting the flow of data from a sender to a receiver to ensure that the receiver can handle all of the incoming data. The most basic flow-control mechanism is dropping. If the resource is not ready (e.g. the buffer is full or

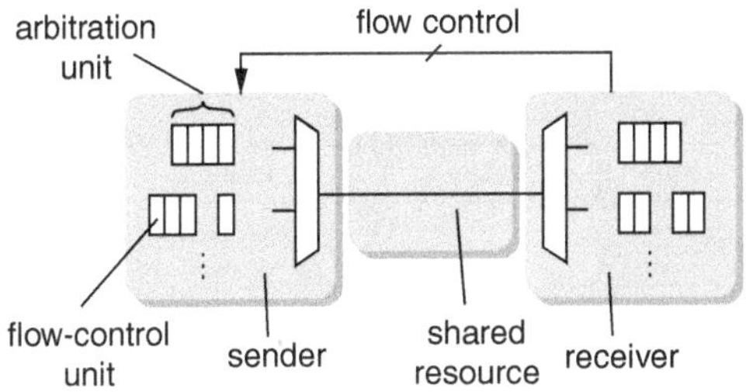

Fig. 2.6 Resource sharing

the receiver is busy), then data is simply dropped. However, dropping data and thus making the flow-control lossy complicates the use of the resource in a larger context and requires expensive recovery mechanisms to provide lossless and ordered service on higher levels [164]. Moreover, as discussed in Section 2.5, lossy flow control conflicts with predictability, as the number of retransmissions must be bounded.

Rather than dropping data, we choose to use a robust flow-control scheme where the producer waits for the availability of the resource, thus avoiding retransmissions. The most basic implementation of lossless flow-control uses (synchronous) handshake signals (e.g. valid and accept) between the sender and receiver. This solution requires no extra book keeping, but restricts the throughput (when the handshake is pipelined) or clock speed (when not pipelined) [150, 158]. An alternative flow-control mechanism uses credits to conservatively track the availability of space on the receiving side. Credit-based flow control can be pipelined efficiently and does not limit the throughput. It does, however, introduce additional counters (for storing credits) and wires (for transmitting credits). As we shall see, our interconnect uses both handshake-based and credit-based flow control at appropriate points.

With the introduction of flow control (handshake or credit based), we can transmit data between the sender and receiver in a lossless fashion. In case the receiver is busy, the sender blocks. For a resource that is not shared, the blocking is not a problem (from the perspective of composability), as only the unique application is affected. However, blocking is problematic for composable sharing as we must ensure that applications do not interfere temporally. That leads us to the next issue, namely the respective granularities of flow control and arbitration.

2.4.2 Flow Control and Arbitration Granularities

The second property that is important for composable resource sharing are the respective flow control and arbitration granularities. To illustrate the problem, consider the initiator bus where the arbiter schedules bus transactions, i.e. complete read and write transactions. Thus, arbitration decisions are taken at the level of transactions. Flow control, however, is done at the finer element level [8, 49, 160]. It is thus possible that a decision is taken and subsequently blocks while in progress, due to lack of either data or space. This is illustrated in Fig. 2.7, which shows the arbitration

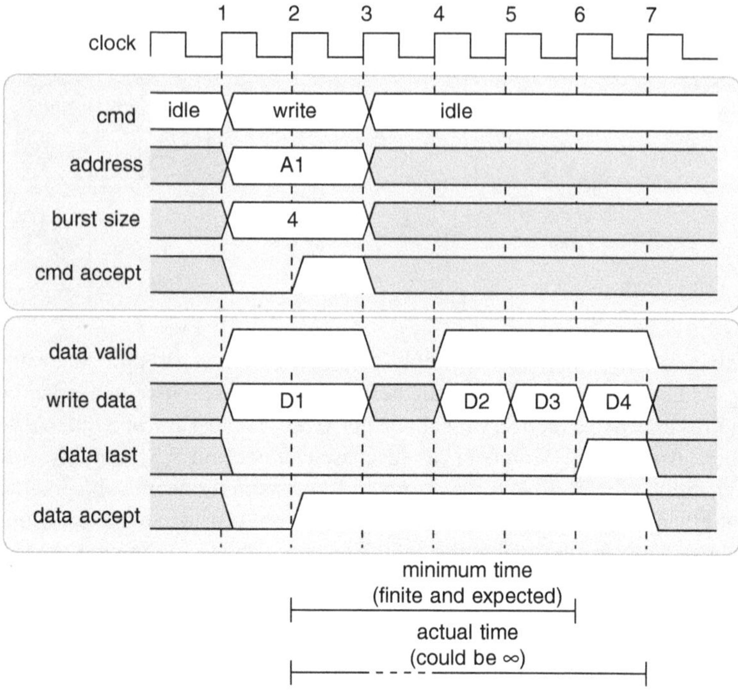

Fig. 2.7 Timing diagram of memory-mapped write transaction

unit of size 4 in Fig. 2.6 in greater detail. Using e.g. DTL [49], a memory-mapped write command (of size 4) is presented in cycle 1, and accepted in cycle 2 together with the first data element. The second data element, however, is not yet available in cycle 3, e.g. due to a slow initiator. Thus, the resource is stalled (and blocked) *for everyone*, waiting for the write data element that is only presented in cycle 4. A similar situation arises for a read transaction when the read data elements are not presented or accepted in consecutive cycles. While the handshakes on the element level ensure lossless transfer of the write data, they also lead to an unknown time before a new arbitration unit can be scheduled, because it depends on the behaviour of the users (IPs and applications) rather than the behaviour of the resource. Hence, due to the discrepancy in levels between taking and executing decisions, the user behaviour affects the resource sharing.

To avoid blocking on the level of flow-control units, we incorporate the availability of data at the source and space at the destination as preconditions for the arbiter in both the NI and the initiator bus. Thus, the arbiter must *conservatively* know how much data is available at the source and how much space is available at the destination and ensure that sufficient data and space are available prior to the scheduling of a transaction or packet. This way, we implement non-blocking flow control on *complete arbitration units*, similar to what is proposed in [132]. As a result, buffering per application is needed both before and after the shared resource. In our case the buffering is provided by the sending (data) and receiving (space) NI for the network, and by the atomiser (data and space) for a shared target.

An alternative to raising the level of flow control to complete arbitration units, as we do in the proposed interconnect, is to change the level of arbitration, and implement *pre-emption* of transactions and packets, respectively. Doing so is nothing more than replacing the shared resource in Fig. 2.6 with one or more instances of the whole figure, as illustrated in Fig. 2.8. In other words, the problem is pushed one level down, and it becomes necessary to offer parallel states, with independent flow control and buffering, for the elements and flits (rather than transactions and packets) of all applications. This is, for example, the approach taken in multi-threaded OCP [147], and in networks based on *virtual circuits* [12, 27, 154, 162, 165] where every router is a resource as depicted in Fig. 2.8.

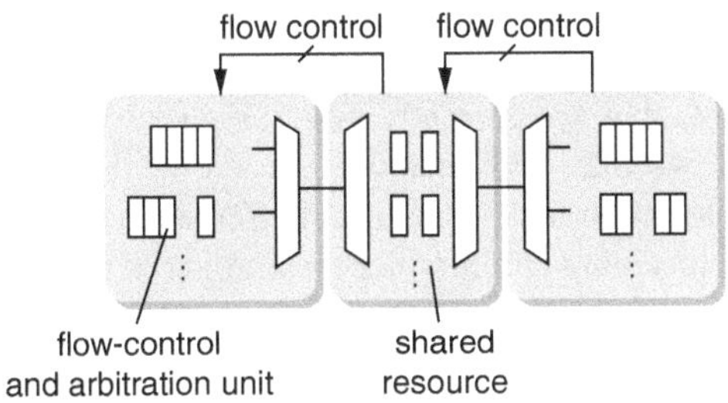

Fig. 2.8 Resource sharing by pre-emption

The reason we choose to raise the level of flow control rather than lowering the level of arbitration for shared targets is that many IPs simply do not support it. Only the OCP protocol offers such support,[3] and even for IPs using this particular protocol it is an optional profile implemented by few IP providers. For the network, the decision is based on design complexity and implementation cost. In contrast to network with virtual circuits, our routers and link pipeline stages are *stateless* [186]. As a result, they are not negatively affected by the number of resource users, in this case the number of applications (translating to a number of virtual circuits, i.e. parallel states and buffers) or the real-time requirements of the applications (the depth of the buffers). Instead, the entire router network (with pipeline stages) in our proposed interconnect can be seen as a single non-blocking shared resource, as shown in Fig. 2.6. The price we pay for the raised level of arbitration is that the NIs require end-to-end flow control, as discussed later in Chapter 3. Moreover, the NI has to know the size of the arbitration units, i.e. the packets. The arbitration unit size is indeed the third important point in enabling composable resource sharing.

[3] AXI currently does not allow multi-threading in combination with blocking flow control.

2.4.3 Arbitration Unit Size

The third point arises due to the choice of raising the flow-control granularity rather than lowering the arbitration granularity (i.e. pre-empting). As a consequence of this decision, the size of an arbitration unit must be known. Moreover, when taking a scheduling decision, the whole arbitration unit must be present in the sending buffer (to avoid placing assumptions on the incoming data rate, i.e. the module inserting data in the buffer). Thus, the maximum arbitration unit size must be known, and the sending buffer sized at least to this size. The same holds for the receiving buffer (with similar arguments as for the sending buffer). In fact, due to the response time of the shared resource and the time required to return flow control, the receiving buffer needs to hold more than one maximum sized arbitration unit to not negatively affect the throughput. We continue by looking at how the problem of variable arbitration unit size is addressed in the network and for the shared targets, respectively.

In the network, we solve the problem with variable packet sizes by letting the arbiter in the sending NI *dynamically adapt the size* of packets to fit with the number of flow-control credits that are currently available. That is, if space is only available for a packet of two flits, then that size is used for the current packet, even if more data is available in the input (sending) buffer. As a consequence, a packet is never larger than the receiving buffer (since the amount of flow-control credits never exceed its size). The buffers can hence be sized in any way (for example to satisfy a specific temporal behaviour as shown in Chapter 6), and the packet sizes adapt dynamically.

For a shared target the situation is more complicated as the size of the arbitration units is decided by the initiators that present the transactions to the initiator bus. Making worst-case assumptions about the maximum arbitration unit is costly in terms of buffering if that size is seldom (or never) used. Moreover, some memory-mapped protocols, most notably AHB [2] and OCP [147], have sequential burst modes where there is no maximum size (referred to as an un-precise burst in OCP). The atomiser, described in detail in Chapter 3, addresses those issues by chopping up transactions into fixed-size sub-transactions that are presented to the arbiter in the initiator bus. The fixed size is also important for the fourth and last item, namely the temporal interference between two arbitration units.

2.4.4 Temporal Interference

The fourth and last part of temporally composable resource sharing is time itself. Composability requires that the time at which an arbitration unit is scheduled and the time at which it finishes does not depend on the presence or absence of other applications. This can be ensured by always *enforcing the maximum temporal interference of other applications*. Note that the response time of the shared resource, i.e. the time it takes to serve a complete arbitration unit, does not matter for composability (although it is important for predictability as discussed later). For example, if we would clock gate the resource (and arbiter) in Fig. 2.6 an arbitrary period of time, this changes the time at which arbitration units are scheduled (unpredictably), but the interference is unaffected and the resource is still composable. However, *enforcing the worst-case temporal behaviour* (per application), also including the uncertainty of the platform, is a sufficient but not necessary condition for composability.

In the network, the maximum interference between applications is enforced by time multiplexing packet injection according to the slot table and enforcing the maximum size (per packet and flit). Even when a packet does not occupy complete flits (i.e. finishes early), the size specified in the slot table is enforced from the perspective of other applications. The links accept one phit every cycle, and there is no contention in the router network (due to the contention-free routing). Consequently, the time between the scheduling of packets is known and determined by the slot table.

For a shared target port, the time between arbitration decisions depends on the transaction granularity of the atomisers (a write requires at least as many cycles as write data elements) and the behaviour of the specific IP. For the SRAM in Fig. 2.1, for example, the atomisers issue transactions of one element, and a new transaction can be issued every two cycles. Similar to the network, the initiator bus uses TDM-based arbitration to enforce the maximum interference between applications.

2.4.5 Summary

To summarise, from the perspective of the applications, our interconnect implements composable resource sharing by using pre-emptive arbitration and enforcing maximum application interference. Thus, *composability is achieved without enforcing the worst-case temporal behaviour*. The network is pre-emptive at the level of connections, but implemented using non-pre-emptive sharing at the level of packets. Similarly, a shared target is pre-emptive at the level of transactions, but implemented using non-pre-emptive sharing at the level of sub-transactions. For this purpose, a new component, namely the atomiser, is introduced. The rationale for not choosing a lower level of pre-emption is reduced cost and compliance with existing memory-mapped protocols. The rationale for not choosing a higher level of pre-emption, i.e. consider the entire interconnect as a non-blocking shared resource, as proposed in the time-triggered architectures [100, 157], is that this pushes the responsibilities of ensuring availability of data and space onto the IPs. In doing so, the interconnect is no longer suitable for general applications, which conflicts with our requirements on diversity.

2.5 Predictability

There are two important parts to predictability, namely having an architecture built from blocks that deliver bounds on their temporal behaviour [12, 26, 63, 95, 108, 122, 154, 157, 165, 193], and choosing a model in which those behaviours can analysed together with the behaviours of the applications [24, 182, 185]. We start by presenting the architecture, followed by the rationale behind the selected analysis technique.

2.5.1 Architecture Behaviour

To provide bounds on the end-to-end temporal behaviour of an application, all the resources used by the application, e.g. the network, must be predictable, i.e. offer useful bounds on their individual temporal behaviours. If a resource is also shared between tasks of the same application, an intra-application arbiter is required to facilitate admission control, resource reservation and budget enforcement [120]. The intra-application arbiter thus prevents a misbehaving or ill-characterised task from invalidating another task's bounds. In contrast to the composable sharing discussed in Section 2.4, the work-conserving [205] intra-application arbiter does not enforce worst-case interference and can distribute residual capacity (slack) freely between the individual tasks (or connections) of the application, possibly improving performance.

In situations where an application requires bounds on its temporal behaviour and a (predictable) resource is shared also with other applications, it is possible to separate intra- and inter-application arbitration. As already mentioned, the intra-application arbiter can distribute capacity within an application and thus reduce the discretisation effects on resource requirements or use the scheduling freedom to improve average-case performance [71]. The cost, however, is in the additional level of arbitration and buffering that must be added. Therefore, in our proposed interconnect, we currently use only one level of arbitration, both within and between applications.

2.5.2 Modelling and Analysis

For application-level predictability, the entire application and platform must be captured in a specific MoC. Thus, a model is needed for every block in Fig. 2.1 that is to be used by a real-time application. Moreover, for the shared resources, i.e. the network and the initiator buses, the arbitration mechanism must be modelled. Having temporal bounds on the behaviour of every component is necessary, but not sufficient. For example, in Chapter 4, we allocate network resources to guarantee the satisfaction of *local latency and throughput bounds*, inside the network. However, as we have seen in Section 2.4, the flow control and arbitration is affected by the availability of buffer space. Consequently, it is necessary to also model the effects of buffers and flow control, between every pair of components, also considering potential variations in the granularity of arbitration units (e.g. read and write transactions).

Taking these considerations into account, we choose to capture the temporal behaviour of the interconnect using Cyclo-Static Dataflow (CSDF) as the MoC [24]. The rationale for doing so is that dataflow analysis [182], in contrast to e.g. real-time calculus, enables us to capture the effects of flow control (bounded buffers) and arbitration in a straightforward manner. As we shall see in Chapter 6, run-time arbiters from the broad class of latency-rate servers [185] can be modelled using dataflow graphs [198]. This enables us to construct a conservative model of a network channel, capturing both the interconnect architecture and resource allocations, derived in Chapter 4. Additionally, dataflow graphs cover a wide range of application behaviours, even with variable-production and consumption rates [199], enabling us to capture the platform and the mapping decisions with a good accuracy, i.e. with tight bounds [127]. Dataflow graphs also decouple the modelling technique and analysis method, thereby enabling us to analyse the same dataflow model with both fast approximation algorithms [15] and exhaustive back-tracking [44]. Using dataflow analysis it is possible to compute sufficient buffer capacities given a throughput (or latency) constraint, and to guarantee satisfaction of latency and throughput (and periodicity) constraints with given buffer sizes [10].

2.6 Reconfigurability

We have already seen examples in Fig. 1.6 of how applications are started and stopped at run time, creating many different use-cases. The different use-cases have different communication patterns (connection topology) and throughput and latency requirements (connection behaviour) that the interconnect must accommodate. Moreover, as stated in Chapter 1, starting or stopping one application should not affect the other applications that are running. We now look at the impact those requirements have on the granularity of reconfiguration and the interconnect architecture.

2.6.1 Spatial and Temporal Granularity

Figure 2.9 shows the example system, with the same progression of use-cases as already shown in Fig. 1.6b, but here illustrating the reconfiguration (resource allocation) granularity, both spatial and temporal. Given that the connection topologies and behaviours vary per use-case, a first approach is to allocate resources per use-case [138], as illustrated in Fig. 2.9a. However, applications are disrupted on use-case transitions, i.e. no composable or predictable services can be provided. Even if an application is present in multiple use-cases, e.g. the filter in the figure, resources used to provide the requested service are potentially different. As the label *global reconfiguration* in Fig. 2.9a illustrates, a use-case transition involves closing and subsequently opening all connections (of all applications) for the two use-cases. Not only does this cause a disruption in delivered services, but it leads to

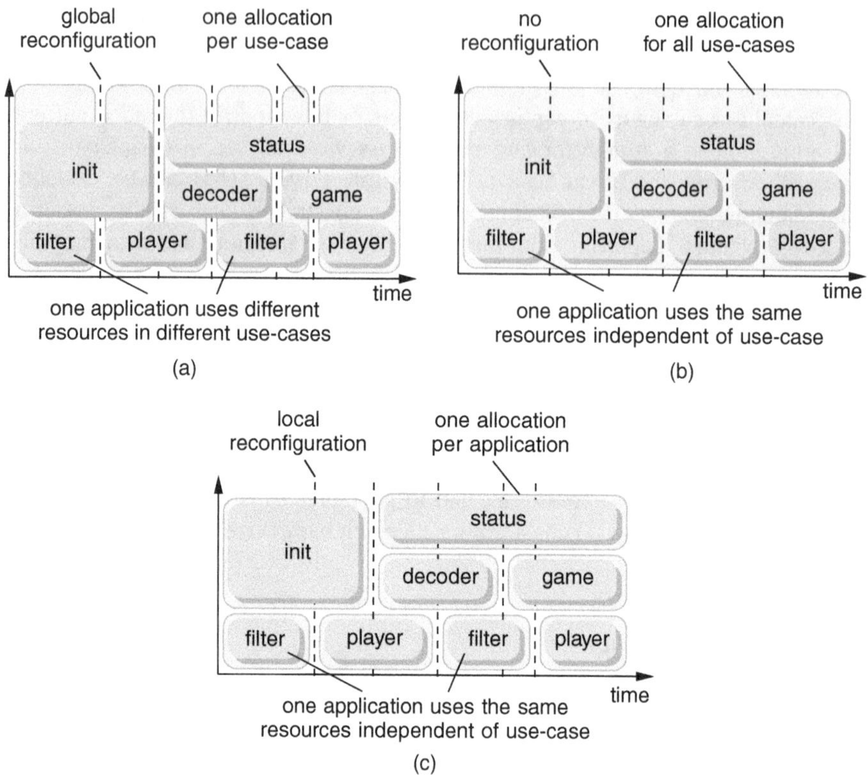

Fig. 2.9 Spatial and temporal allocation granularities

unpredictable reconfiguration times as in-flight transactions of all the running applications must be allowed to finish [102, 145], even if we only want to reconfigure one application.

Undisrupted service for the applications that keep running is achieved by extending the temporal granularity to one allocation for all use-cases [137]. As shown in Fig. 2.9b this removes the need for reconfiguration on use-case transitions. While delivering undisrupted services to the applications, the requirements of the synthetic worst-case use-case are over-specified, with a costly interconnect design as the result [138]. Moreover, if the architecture of the interconnect is given, it may not be possible to find an allocation that meets the worst-case use-case requirements, even though allocations exist for each use-case individually. The worst-case approach is also not applicable to connections that make direct use of the streaming stack. Since a streaming port is only connected to one other port at any given point in time, reconfiguration is required.

The approach we take in this work is to perform resource allocation on the granularity of applications, as later described in Chapter 4. The rationale for doing so is that it introduces spatial (in addition to temporal) granularity thus allowing *local reconfiguration*, as shown in Fig. 2.9c. Similar to the worst-case approach, applications are allocated with the same resources independent of the use-case and are unaffected when other applications are started or stopped. However, reconfiguration enables us to share resources (bus ports, NI ports, links and time slots) between mutually exclusive applications (e.g. init, decoder and game in Fig. 2.9c), thus reducing the cost of the interconnect [72].

Although our proposed approach allows the user to specify both a spatial and temporal granularity of reconfiguration, it is not required to use this feature, i.e. it is possible to only distinguish the temporal axis. This is achieved by having one single worst-case application (a fully connected constraint graph), or mutually exclusive applications (no edges in the constraint graph). The methodologies in [137, 138] are thus subsumed in this more general framework.

2.6.2 Architectural Support

There are three modules in the interconnect architecture that are reconfigurable: the target bus, NI and initiator bus. The rationale behind making the target bus and the NI reconfigurable is that they are the two modules where different destinations are decided upon. In the target bus, an initiator port is selected based on the destination address. Similarly, in the (source) NI, a streaming initiator port (in the destination NI) is selected based on the path and port identifier. The NI is also one of the two locations where arbitration is performed. Thus, both the NI and initiator bus are reconfigurable, such that the service given to different connections can be modified at run time.

In Fig. 2.1 we see that the aforementioned modules have a memory-mapped control port. Thus, all reconfiguration is done using memory-mapped communication, reading and writing to the control registers of the individual blocks. The rationale for using memory-mapped communication is the diversity it enables in the implementation of the host. Any processor that has a memory-mapped interface can be used in combination with the run-time libraries, as we shall see in Chapter 5.

2.7 Automation

The proposed design flow, as shown in Fig. 1.8, extends on the flow in [62] and addresses two key problems in SoC design. First, the need for tools to quickly and efficiently generate application-specific interconnect instances for multiple applications. Second, the need for performance verification for a heterogeneous mix of firm, soft and non-real-time applications. We first discuss the rationale behind the input and output of the flow, followed by our reasons for dividing the design flow into multiple independent tools.

2.7.1 Input and Output

As we have already discussed in Chapter 1, this work takes as its starting point the specification of the physical interfaces that are used by the IPs, constraints on how the applications can be combined temporally, and the requirements that each application has on the interconnect. The reason for focusing on the requirements on the interconnect rather than application-level requirements is to allow application diversity by *not placing any assumptions on the applications*. The real-time requirements are described per application, and on a level that is understood by the application designer, e.g. through latency and throughput bounds. Furthermore, the design flow makes no distinction whether the requested throughput and latency reflect the worst-case or average-case use of a particular application. Thus, depending on the applications, the input to our design flow might stem from analytical models, measurements from simulation models, or simply guesstimates based on back-of-the-envelope calculations or previous designs.

Based on the input specification, it is the responsibility of the design flow to automatically generate a complete hardware and software interconnect architecture. There are two important properties of the design-flow output. First, it is possible to turn the output into a physical implementation in the form of an FPGA or ASIC. Second, depending on the application, it is possible to verify application-level performance using a variety of simulation-based and formal techniques. Hence, also the output of the design flow reflects the application diversity.

2.7.2 Division into Tools

The design flow is split into separate tools for several reasons. First, breaking the design flow in smaller steps simplifies steering or overriding heuristics used in each of the individual tools, enhancing user control. It is, for example, possible for the user to construct the entire architecture manually, or modify an automatically generated architecture by, e.g., inserting additional link-pipeline stages on long links. Second, splitting the flow reduces the complexity of the optimisation problem, and simpler, faster heuristics can be used. This results in a low time-complexity at the expense of optimality. As we shall see, pushing decisions to compile time is key as it enables us to guarantee, at compile time, that all application requirements are satisfied, and that all the use-cases fit on the given hardware resources. However, complexity is moved from the platform hardware and software to the design tools, leaving us with a set of challenging problems. As we shall see in Chapters 4 and 6, tools that rely on low-complexity approximation algorithms are central to the interconnect design flow. Higher-level optimisation loops involving multiple tools can then be easily added, as illustrated by the arrows on the left hand side of Fig. 1.8. Third, parts of the flow can be more easily customised, added, or replaced by the user to tailor the flow or improve its performance.

2.8 Conclusions

In this chapter, we present the rationale behind the most important design choices of our interconnect. We introduce the modular building blocks of the hardware and software architecture, and explain how they, together with the design flow, enable us to satisfy the requirements in Chapter 1. In the succeeding chapters, we return to the concepts introduced in this chapter as we explain in-depth how the dimensioning, allocation, instantiation and verification contribute to the requirements.

Chapter 3
Dimensioning

Composable and predictable services require allocation of resources. Prior to the allocation, however, the resources must be dimensioned. Additionally, the resources must enable an allocation to be instantiated and enforced. In this chapter, which corresponds to Step 1 in Fig. 1.8, we show how the architectural building blocks are dimensioned and how they implement the aforementioned requirements.

The modules of the interconnect are introduced top down, following the steps in Fig. 3.1. Step 1.1 bridges between the network and the IP ports by dimensioning buses (Section 3.1) together with atomisers (Section 3.2), protocol shells (Section 3.3) and clock domain crossings (Section 3.4). Thereafter, Step 1.2 dimensions the network topology, namely the NIs (Section 3.5), the routers (Section 3.6) and links (Section 3.7). The final part of the interconnect dimensioning is the addition of the control infrastructure (Section 3.8), corresponding to Step 1.3 in Fig. 3.1. All the modules presented in this chapter, i.e. buses, atomisers, shells, clock domain crossings, NIs, routers and link pipeline stages, are available both as SystemC models and synthesisable HDL, demonstrated on several FPGA instances of the platform. We conclude this chapter by summarising how the dimensioning contributes to the requirements from Chapter 1 (Section 3.9).

3.1 Local Buses

Distributed memory communication is implemented by the *target bus*, as described in Section 3.1.1. The target bus is complemented by the *initiator bus* that implements shared-memory communication, as elaborated on in Section 3.1.2. Next, we describe the target bus and initiator bus in more detail.

3.1.1 Target Bus

A target bus, as shown in Fig. 3.2a, connects one memory-mapped initiator to multiple targets. The target bus is multiplexer based and very similar to an AHB-Lite layer [2]. The primary responsibility of the target bus is to direct requests to the

A. Hansson, K. Goossens, *On-Chip Interconnect with Aelite*, Embedded Systems, DOI 10.1007/978-1-4419-6865-4_3,

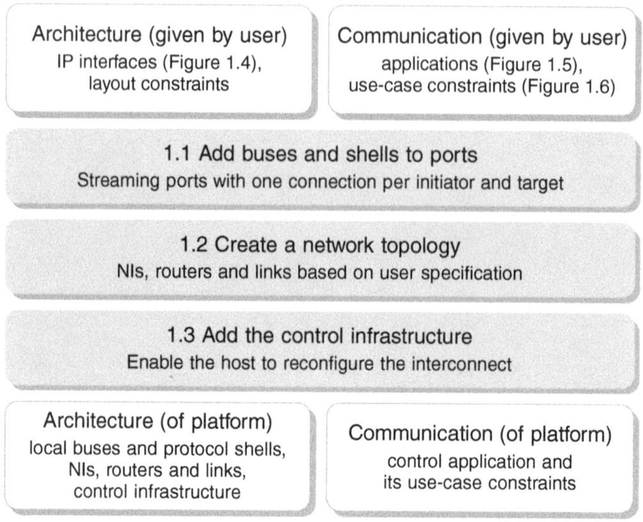

Fig. 3.1 Dimensioning flow

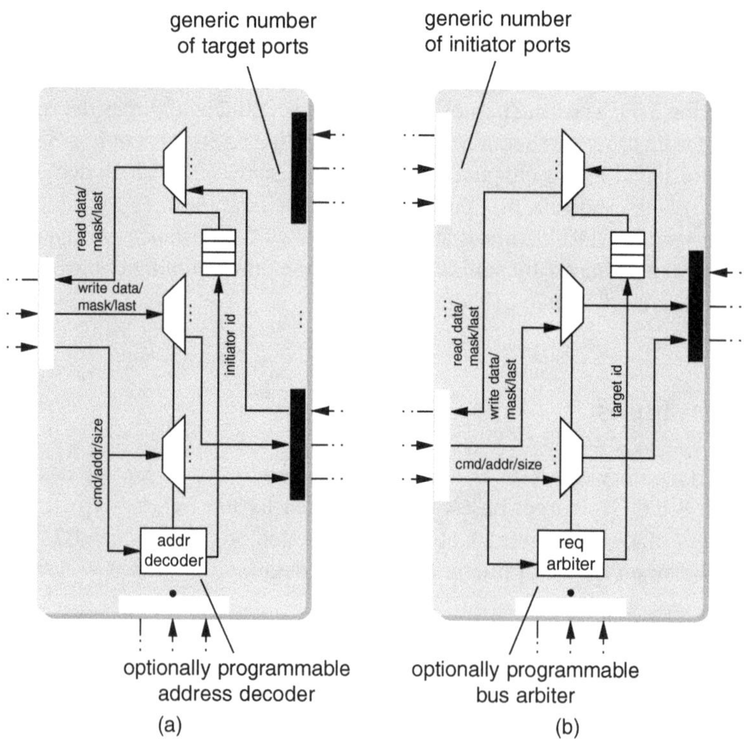

Fig. 3.2 Local target bus (**a**) and initiator bus (**b**) architectures

appropriate target, based on the address of the request. To reduce the negative impact of latency, the target bus allows multiple outstanding transactions, even to different targets. The target bus also enforces response ordering according to the protocol specification. That is, responses are returned in the order the requests where issued (within a thread [8, 147], if applicable). As seen in Fig. 3.2a, the ordering of responses is enforced by storing a target identifier for every issued request. These identifiers are then used to control the demultiplexing of responses. The ordering guarantees of the target bus, together with mechanisms like tagging [49], are leveraged by the IP to implement a certain memory-consistency model, e.g. release consistency, thus enabling the programmer to reason about the order in which reads and writes to the different targets take place.

As exemplified by the ARM and the host in Fig. 2.1, a target bus is directly connected to all initiator ports that use distributed memory communication. Each target bus is individually dimensioned by determining the number of concurrent targets accessed by the initiator it is connected to. Traditional bus-based systems require the designer to determine *which targets* should be reachable (if using sparse bus-layers), and what static address map to use. We only have to determine *how many targets* should be reachable and not which ones. At run time, the target bus address decoder is reconfigured through the memory-mapped control port, as illustrated by the control port in Fig. 3.2a. Thus, the address map is determined locally per target bus, and per use-case. As we shall see in Section 3.8, the target bus is also used in the control infrastructure, but then with a static address decoder.

3.1.1.1 Experimental Results

Figure 3.3 shows the synthesis results of the target bus as the number of initiator ports is varied. Two different bus instantiations are evaluated, with a programmable and fixed address decoder, respectively. In both cases, we limit the address decoding to the five topmost bits of the 32-bit address and allow a maximum of four outstanding responses (stored in a fully synchronous pointer-based flip-flop FIFO). The width of the read and write data interfaces are 32-bits.

Throughout this work, synthesis results are obtained using Cadence Ambit with NXP's 90-nm Low-Power technology libraries. For a given clock frequency target, we use a 50% clock duty cycle, 20% of the cycle time as input and output delay with 10% clock skew. We disable clock-gate insertion as well as scan insertion and synthesise under worst-case commercial conditions. For every design point, e.g. for the different number of ports in Fig. 3.3, we perform a binary search on the target frequency to establish the maximum achievable frequency, and stop when the difference in achieved cycle time is less than 0.1 ns. Note that all synthesis results reported throughout this work are *before place-and-route*, and include *cell area only*. After layout, the area increases and the maximum frequency drops (an utilisation higher than 85% is difficult to achieve and frequency reductions of up to 30% are reported in [162] for a 65-nm technology).

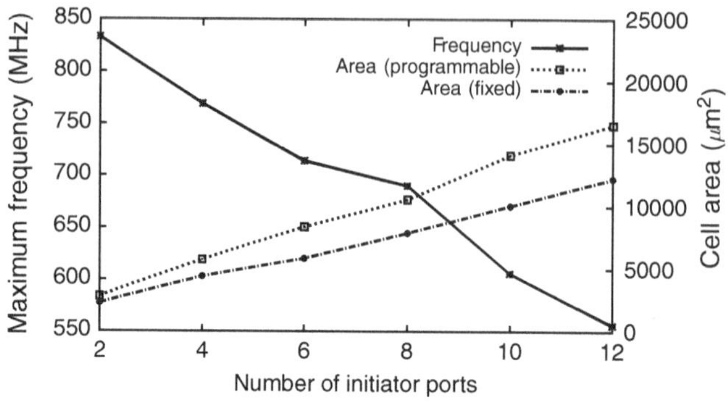

Fig. 3.3 Target bus area and frequency

The first thing to note about the results in Fig. 3.3 is that the maximum frequency for the two architectures is the same, for any number of ports. This is due to the fact that the address decoding is pipelined in both cases, and the critical path starts at the registered decoder output. Even with 12 ports, the target bus runs at more than 550 MHz, which is sufficient for most contemporary IPs. The cell area for the target bus is minuscule, in the order of 0.01 mm^2, with the non-programmable bus occupying slightly less area than the programmable one.

3.1.1.2 Limitations

Currently, the only protocol supported by the HDL libraries is DTL. As DTL is single-threaded, read data and write (tag) acknowledgements are returned to the initiator in the order the requests where issued. This is fine for simple applications that expect responses from different target to come back in order, but it also couples the behaviour of multiple applications mapped to the same IP. In such cases, the initiator port becomes a shared resource, much like what is described in Section 2.4. Thus, a similar mechanism is required to enable composable sharing of the port. In our current implementation we *do not allow multiple applications to share an initiator port*. Moreover, our implementation of tagging [49] only supports tags that are issued concurrently with a write transaction (with write valid driven high), and not retroactive tagging, where the request has already finished. This is not inherent limitation and can be addressed in future implementations of the bus.

3.1.2 Initiator Bus

An initiator bus, as shown in Fig. 3.2b, connects multiple memory-mapped initiators to a shared target. The initiator bus is responsible for demultiplexing and multiplexing of requests and responses, respectively. Similar to the target bus, the initiator bus implements transaction pipelining with in-order responses. It is the primary responsibility of the initiator bus, together with the atomisers, as we shall see in

Section 3.2, to *provide composable and predictable sharing* of the target port. The only requirement placed on the application designer is that *shared target locking* [8] (depreciated in the AXI standard) is not used. Alternatively locks could be used per application, based for example on memory regions. Note that the arbitration in the initiator buses is decoupled from the arbitration in the network, and that different initiator buses can have different arbiters with varying characteristics.

Initiator buses are placed in front of shared target ports, as exemplified by the SRAM and peripheral in Fig. 2.1. Similar to the target buses, each initiator bus is dimensioned individually based on the maximum number of concurrent initiators sharing the target port. If the target port is only shared within an application, we instantiate a predictable (work-conserving) round-robin arbiter. Should the port be shared between applications, which is the case for both the peripheral port and SRAM in our example system, we instantiate a composable and predictable TDM arbiter. Similar to the target buses, the initiator buses are reconfigured at run time using memory-mapped control ports. Depending on the arbiter, the bus is run-time configured with, e.g., an assignment to TDM slots, the size of a TDM wheel, or budget assignments for more elaborate arbiters [3].

3.1.2.1 Experimental Results

The synthesis results in Fig. 3.4 show how an initiator bus with a non-programmable round-robin arbiter (the most complex arbiter available in the current implementation) and a maximum of four outstanding transactions scales with the number of target ports.[1] Similar to the target bus, the total cell area is in the order of 0.01 mm^2, even when the operating frequency is pushed to the maximum. Moreover, even the larger instances of the bus run at more than 550 MHz, which is more than what most IPs require. The initiator bus can be pipelined further if higher frequencies are desired.

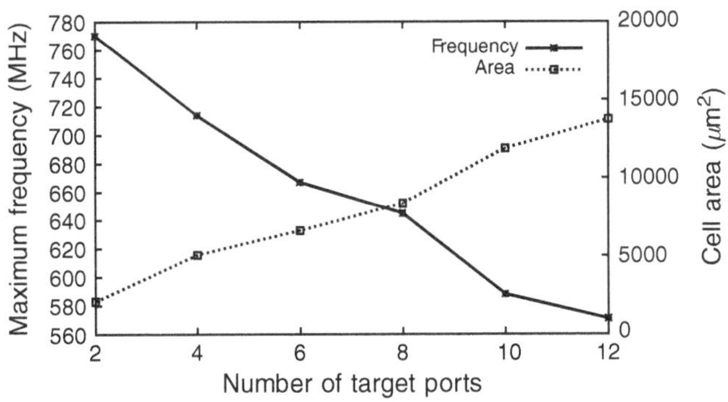

Fig. 3.4 Initiator bus area and frequency

[1] Note that the outstanding transactions are only for the target itself. There could be many more ongoing transactions in the interconnect.

3.1.2.2 Limitations

Similar to the initiator bus, the only protocol currently supported by the HDL libraries is DTL. Moreover, the bus arbitrates at the transaction level, but performs flow control at the element level. Thus, as discussed in Section 2.4, two challenges remain to share a target composably and predictably. First, the level flow control must be raised to the level of arbitration, i.e. complete transactions. Second, the transaction size must be bounded. Hence, the initiator bus needs to be complemented with atomisers to provide composable and predictable sharing of the target port.

3.2 Atomisers

The atomiser, shown in Fig. 3.5, complements the initiator bus by breaking multi-word transactions into multiple one-word transactions, and by performing flow control at the level of transactions. As seen in the figure, the atomiser transparently splits and merges transactions from the perspective of the initiator using it. This is achieved by storing the original transaction sizes and merging responses, e.g. by only asserting the read-last signal when a read transaction completes. After the split, the uniformly sized transactions are passed on to the request generator, before being presented to the arbiter in the initiator bus.

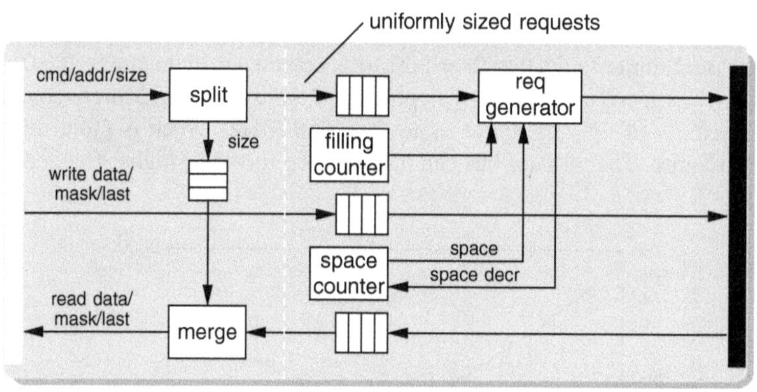

Fig. 3.5 Atomiser architecture

The request generator implements transaction-level flow control, similar to the Memory Transaction Level (MTL) protocol [132]. A write request is only presented to the target bus when all request data is present, i.e. the filling of the request buffer is larger than the burst size (in the current implementation fixed at one element). Similarly, a read request is only presented when the response data can be immediately accepted, i.e. there is sufficient space in the response buffer. The space counter is decremented already at the moment the request is presented on the initiator port. This is to ensure that the buffer space is available at the time the response data arrives, independent of any pipelining in the target the atomiser is connected to.

As part of the dimensioning, an atomiser is added before every port of an initiator bus, thus enabling pre-emptive scheduling and flow control at the level of transactions. Together with appropriate arbiter in the initiator bus (e.g. a TDM arbiter), the atomisers provide composable and predictable sharing of the target port. The atomiser achieves a frequency of 650 MHz and occupies an area in the order of 0.01 mm^2 (dominated by the response buffer) when dimensioned for an SRAM with six cycles response time.

3.2.1 Limitations

The interleaving of transactions enabled by the atomiser potentially violates the atomicity of the protocol used by the IPs. However, for AXI [8], OCP [147] and DTL [49] atomicity is only assumed at the byte level, which our current implementation complies with. Alternatively, in multi-threaded protocols [8, 147], the parallelism between connections can be made explicit via OCP connection- and AXI thread-identifiers.

An important limitation of the proposed atomiser architecture is that it is restricted to (sub-)transactions of a single word. This fits nicely with the SRAM available in our experimental platform, as we shall see in Chapter 7. SDRAM controllers, however, typically require larger bursts to achieve a reasonable memory efficiency [3]. We consider it future work to add shared SDRAM, and thus also more elaborate atomisers, to the platform template.

3.3 Protocol Shells

The protocol shells bridge between memory-mapped ports and the streaming ports of the network. As seen in Fig. 2.1, shells are connected either directly to the IPs (on the μBlaze, VLIW and video tile), to the ports of the target buses (on the bus of the host and ARM), or to the atomisers of the initiator buses (on the bus of the peripheral and SRAM). For a specific memory-mapped protocol there is a target shell, an initiator shell and their associated message formats, as shown in Fig. 3.6. Next we give an overview of the functionality of the shells, followed by an in-depth discussion of their architecture.

To illustrate the functionality of the shells, consider the DTL target shell in Fig. 3.6a. The request encoder awaits a valid command (from the initiator port connected to the shell), then serialises the command, burst size, address and potential flags. In the case of a write, it also serialises the data elements, together with the masks (sometimes referred to as byte enables or strobes) and the write last signal. In the initiator shell, the request decoder does the opposite, and on the arrival of a request message it drives the command, address and burst size signals. If the transaction is a write, the initiator shell also presents the write data along with its masks and the last signal. Responses follow a similar approach in the opposite direction.

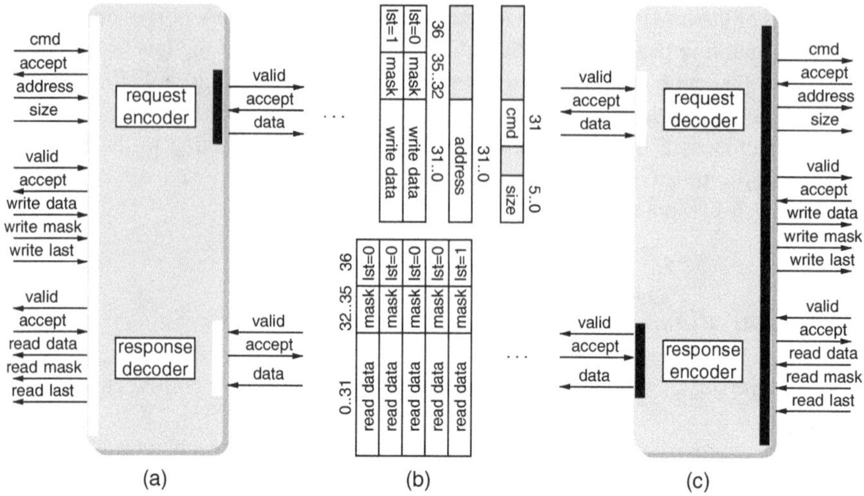

Fig. 3.6 Target shell (**a**) and initiator shell (**c**), with associated message formats (**b**)

The initiator shell encodes the read data, together with the mask and last signal, only later to be decoded and reassembled by the target shell.

The shell architecture is largely dependent on three properties of the memory-mapped protocol. First, the division into a request and response part depends on whether the protocol supports split transactions or not. Second, the number of streaming port pairs (connections) depends on the amount of concurrency that the protocol allows, e.g. through independent threads. Third, the message format depends on the signals of the interface. We now discuss these three properties in turn.

The division into an independent request and response part, as shown in Fig. 3.6, is suitable for protocols like DTL, AXI and OCP, where requests and responses are split. Additionally, the write- and read-last signal (or similar) is used by the encoders and decoders to determine the end of requests and responses (as opposed to using counters). Thus, the finite state machines that implement the encoders and decoders are only coupled externally, through the initiator or target they are connected to, but not internally within the shells. For protocols like AHB and OPB, with limited or no support for split transactions, there is a much tighter coupling and the state machines are either merged or communicating, thus placing the AHB protocol shell on the transport level in the protocol stack, as discussed in Section 2.3.

The second important design choice in the shell architecture is the number of pairs of streaming port, i.e. connections. More connections give more parallelism, but do so at a higher cost, and offer little or no benefit if the protocol does not exploit the added parallelism. Thus, for DTL (and PLB), where the command group for read and write is shared, we choose to only use one connection per memory-mapped initiator and target pair. Thus, the shells in Fig. 3.6 have one pair of streaming ports. For protocols like OCP or AXI, with independent read and write channels, two connections can be used, one for read transactions and one for write transactions. Moreover, with support for multiple independent threads [8, 147], each thread can be given its own connection(s). However, our separation of the protocol stacks allows the thread identifier [8] and connection identifier [147] to be used at other granularities than the connections. Hence, the shells enable different amounts of parallelism for different protocols, but for simplicity we assume one connection per shell throughout this work.

The third property that is of great importance is the message format. Normally, the streaming protocol is narrower, i.e. uses fewer wires, than the memory-mapped protocols. Thus, as exemplified by the command group and write data group in Fig. 3.6b, the signal groups of the memory-mapped interfaces are (de)serialised. The proposed message format is tailored for DTL, but is suitable also for other similar protocols like PLB. Different shells may use different message formats, allowing multiple memory-mapped protocols to co-exist. We do not attempt to use a unified message format, as it is typically not possible to seamlessly translate between different protocols (although such claims are made in [7]). The reason is that the protocols are not compatible on the higher levels in the protocol stack, e.g. in terms of ordering, errors, masking, etc.

The DTL initiator and target shell are not parametrisable, and achieve a maximum frequency of 833 MHz, with an area of 2,606 and 2,563 μm^2, for initiator shell and target shell, respectively.

3.3.1 Limitations

The current implementation of the shells assume that the address and data width of the memory-mapped interface is 32 bits. Additionally, our current implementations of the shells do not support error codes and aborted transactions. We consider it future work to extend the functionality of the shells. Moreover, due to the serialisation and header insertion, the shells emphasise the importance of burst transactions. The aforementioned issue can be solved by moving the clock domain crossing into the shell (assuming the network is running at a higher clock speed). In our proposed interconnect, the shell is in the IP clock domain, and a separate block is responsible for the clock domain crossing.

3.4 Clock Domain Crossings

For simplicity, and compatibility with both ASIC and FPGA design flows, this work uses a grey-code pointer-based bi-synchronous FIFOs [41] to implement the clock domain crossings between IPs and the network. The FIFOs are compatible with standard CAD tools [103] and let the IPs (or shells) robustly interface with the network with high throughput (one transfer per clock cycle) and a small latency overhead (two clock cycles). As the last part of Step 1.1 of the dimensioning flow

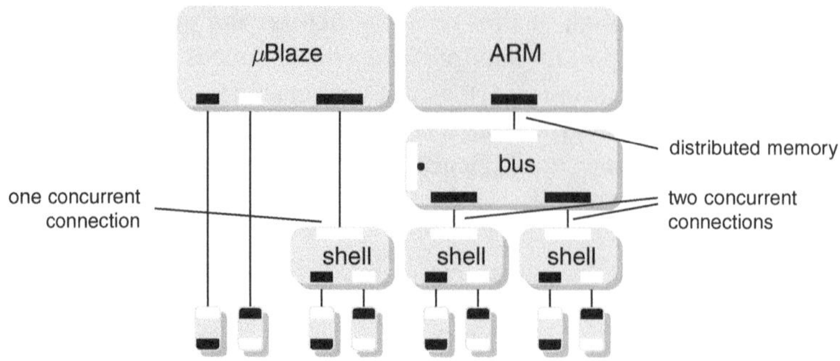

Fig. 3.7 IP port dimensioning

in Fig. 3.1, a bi-synchronous FIFO is connected to all streaming ports on shells or IPs that are not in the network clock domain. Figure 3.7 shows the end result for the μBlaze and ARM, with buses, shells and clock domain crossings added to their ports based on the type of port and number of concurrent connections.

The bi-synchronous FIFO scales to high clock speeds, achieving more than 700 MHz for a 37-bit wide and 3-word deep FIFO using our 90-nm technology libraries. The aforementioned FIFO instance occupies roughly 5,000 μm^2. The same FIFO architecture is used for the flit-synchronous links in Section 3.7. More efficient implementations of bi-synchronous FIFOs, with lower latency and area requirements, are presented in [154, 196]. These FIFOs could be used as part of our interconnect to improve the performance and reduce the cost.

3.5 Network Interfaces

After the addition of the clock domain crossings, we move to Step 1.2 of the dimensioning flow in Fig. 3.1 and create a network topology. The network, consisting of NIs, routers and links, connects the streaming ports on the shells and IPs over logical connections. As already discussed in Chapter 2, a connection comprises two uni-directional channels. The NI is responsible for providing each channel with lossless and ordered data transmission, with an upper bound on latency and a lower bound on throughput.

To avoid buffer overflows *within the network* and thus provide lossless services, *credit-based end-to-end flow control* is used.[2] As we shall see, each channel has a dedicated FIFO in both the sending and receiving NI. To ensure in-order delivery, the path through the network is determined per channel by *source routing*. That is, the path is embedded in the packet header by the sending NI, and the routers follow the directions in the header. The path is thus the same for all flits belonging to a

[2] For best-effort connections in the Æthereal architecture, the end-to-end flow control ensures freedom from message-dependent deadlock [73].

specific channel. The bounds on latency and throughput are provided by means of contention-free routing, as described in Section 2.1. We now look at the details of the NI architecture.

3.5.1 Architecture

Figure 3.8a illustrates the division of the NI into a buffer part a scheduler part, and a register file, and highlights the subcomponents together with their control and data dependencies. We present the subcomponents in turn.

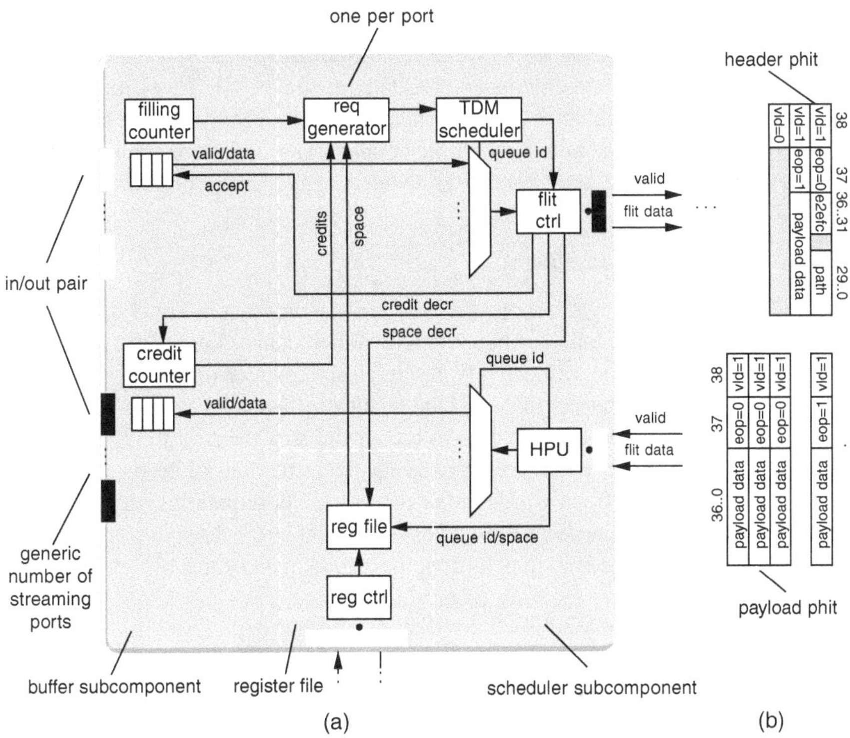

Fig. 3.8 NI architecture (**a**) with associated flit format (**b**)

3.5.1.1 Buffer Subcomponent

The buffer subcomponent is dimensioned for a specific number of streaming port pairs, hereafter referred to simply as a port since they always appear in pairs. As we shall see in Chapter 4, we defer the NI dimensioning to the resource allocation, where also the mapping of IPs to NIs is decided. Thus, in Step 1.2 in Fig. 3.1, the NI is dimensioned with only one streaming port to be used by the control infrastructure as discussed in Section 3.8.

Each input and output port has a unique FIFO. Each target streaming port corresponds to an *input queue*, with data from the IP to the network. Similarly, an initiator FIFO port corresponds to an *output queue*. The size (number of words) and implementation (e.g. asynchronous or synchronous, ripple-through [196] or pointer-based, binary or grey-code pointers [41], state retention or not) is selected per FIFO. The input and output queues are given default sizes (determined on a per-network basis) in the dimensioning flow. In Chapter 6 we show how to determine minimal buffer sizes given the application requirements.

A *filling counter* (conservatively) tracks the occupancy in the input queue.[3] Similarly, each output queue corresponds to a *credit counter*. The latter (conservatively) tracks the number of words freed up in the FIFO that are not yet known to the sending NI. Throughout this work, we assume that the entire NI is in one clock domain, with any clock domain crossing located outside the NI. This simplification (with respect to [168]) enables us to use synchronous counters for queue filling and credits. The number of bits allocated for the counters, uniform throughout the NI, is also a design-time architectural parameter decided by the largest buffer size used.

3.5.1.2 Scheduler Subcomponent

The scheduler subcomponent has one flit port, seen on the right in Fig. 3.8a, and one port for each FIFO in the buffer subcomponent. When data or credits are present in the input and output queue, respectively, the request generator for that port informs the scheduler. Based on the slot table and the architectural constants in Table 3.1, the scheduler decides from which port data or credits are sent the next flit cycle. Thus, scheduling is done at the level of flits, as is the decrementing of credit and space counters, although these track words rather than flits. Consequently, with a flit size s_{flit} of three words, the scheduler only takes a decision every three cycles. Counter updates are thus pipelined and do not affect the critical path of the NI.

Table 3.1 Architectural constants of the NI

Symbol	Description	Value	Unit
s_{flit}	Flit size	3	Words
s_{tbl}	Slot table size	–	Flits
s_{hdr}	Packet header size	1	Words
s_{pkt}	Maximum packet size	4	Flits
s_{crd}	Maximum credits per header	31	Words

The flit control unit is responsible for constructing the actual flit for the port decided upon by the scheduler. If the current port is different from the one scheduled in the previous flit cycle, a *packet header* containing the path and any potential

[3] In the case of a pointer-based FIFO the counter can be implemented as part of the FIFO.

credits is inserted, as shown in Fig. 3.8b. The header occupies s_{hdr} words of the flit. The remaining words of the flit (assuming $s_{\mathrm{hdr}} < s_{\mathrm{flit}}$) carry payload in the form of data from the input queue. Headers are also forcefully inserted after a maximum packet size s_{pkt} to prevent starvation of credits (as discussed later in Chapter 6). The flit control is responsible for updating the space and credit counters, and for driving the accept signal to the input queue as data is sent. As shown in Fig. 3.8b, a fixed number of bits are reserved for sending credits. Consequently, maximally s_{crd} credits can be sent in one header.

In the other direction, for flits coming from the router network, the Header Parsing Unit (HPU) decodes the flits and deliver credits to the appropriate space counter, and data to the appropriate output queue. Note that there is no need for an accept signal from the output queue to the HPU due to the end-to-end flow control.

3.5.1.3 Register File

The register file is programmable through a memory-mapped target port and is divided into three distinct tables: (1) the *channel table* containing the path and a request-generator enable bit; (2) the *space table*, holding the current space counter value, but also the programmed maximum as a reference and two enable bits for the end-to-end flow control; and (3) the *slot table* which contains the time wheel used by the scheduler. The channel and space table have one entry *per target port*, whereas the slot table is of arbitrary length, and holds port identifiers (that are one-hot encoded in a pipelined fashion before they are used as control signals by the multiplexer in front of the flit control). The slot-table size s_{tbl}, and even the register file organisation, is determined per network instance.

The most important property of the register file is the fact that it enables run-time reconfiguration. Moreover, the register file is organised such that one connection (port) can be modified without affecting other connections on the same NI, thus enabling *undisrupted partial reconfiguration*, something we return to discuss in Chapter 4. Note that it is not possible to program the credit counters. In fact, for reconfiguration to be robust, changes to a channel must only take place when those counters are at zero, and there is no data in the FIFOs or in the router network. We discuss these requirements in more depth in Chapter 5. To implement robust reconfiguration the register file offers *observability of quiescence* for the individual ports through the reference space counters and filling counters.

3.5.1.4 Flit Format

The NI schedules connections at the level of flits, each one further subdivided into *physical digits* (phits), i.e. the actual words sent over the wires between two modules. Due to the use of *source routing*, there are two different types of phits: headers and payload. A detailed description of the two types follow.

When a new connection is scheduled, i.e. a new packet starts, the first flit carries a *packet header*, consisting of the path to the destination FIFO and any potential credits belonging to the connection in the destination NI. As already mentioned, throughout this work we assume that the header size, denoted s_{hdr} is one phit. Thus, as seen in Fig. 3.8b, the header (1 phit) does not occupy the entire flit, and even a header flit is able to carry payload data (2 phits). Within the header phit, the path field is 30 bits and holds a sequence of output ports, encoding the path through the router network and lastly the port in the destination NI. Along each hop, the router or NI looks at the lowest bits (corresponding to the 2-logarithm of its arity) of the path and then shifts those bits away. The 30 bits are sufficient to span 10 hops in a network where all the routers and NIs have 8 ports (or less). An example of such a topology is a 5 by 5 mesh with 4 NIs per router, i.e. a total of 100 NIs, supporting roughly 1,000 connections (depending on the number of ports per NI). Not having fixed bit fields, as used in e.g. [12] (where a hop is always represented by three bits), *allows arbitrary topologies* with a varying number of ports per router and NI.

In contrast to all other NoCs we are aware of, we impose no restrictions on the path, i.e. even loops are allowed. Due to the contention-free routing there is no need to impose restrictions in order to avoid routing deadlock or message-dependent deadlock [73]. Consequently, we allow flits to turn around in a router and go back to the output port that corresponds to the input where the flit came from. As we shall see in Chapter 4, this is used when two pairs of streaming ports on the same NI are communicating with each other. The connection then goes from one port on the NI, to the router and back to another port on the same NI.

The payload phits, unlike the header phit, have no meaning to the NIs or routers, and are simply forwarded. In other words, like a true protocol stack, *the network is oblivious to the contents of the payload.* A flit contains no information about its payload, unlike NoCs that rely on Virtual Circuits (VC) to avoid message-dependent deadlock [73], in which case the flits have control information about the messages they carry to enable the network to put them in the appropriate buffers or use specific scheduling algorithms.

The flit format allows *arbitrary-length packets*, but as already discussed, the NIs end packets after a maximum number of flits. The latter is important to provide temporal bounds on the return of credits as we shall see in Chapter 6.

3.5.2 Experimental Results

We split the synthesis of the NI into (1) buffers, (2) slot table and (3) the complement of these parts. The reason for the division is that the buffers grow with their depth, the slot table in the number of ports and the number of slots, and the remainder (scheduler and register file) in the number of ports. A discussion of the three parts follow.

All NI buffers are of a uniform width, 37 bits, corresponding to the message format of the shells, as discussed in Section 3.3. The synthesis results in Fig. 3.9 therefore show the maximum frequency and the associated cell area for 37-bit wide FIFOs of varying depth. The FIFO implementation in question is a fully synchronous, pointer-based Flip-Flop FIFO. We see that the larger FIFOs achieve

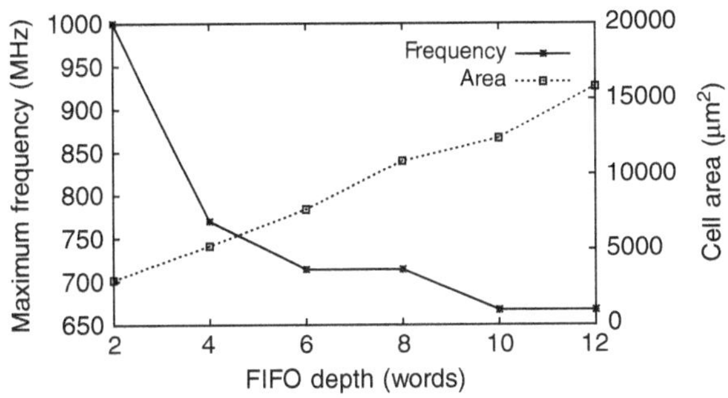

Fig. 3.9 Flip-flop FIFO area and frequency

maximum frequencies of around 650 MHz. The area grows linearly with the depth, as expected. The actual size of the FIFOs depends on the application and buffer sizing, a topic we return to in Chapter 6.

The second part of the NI is the slot table, implemented as a register file. The width of the register is the number of bits needed to represent all the ports (plus one to denote an unscheduled slot). The depth is the number of slots, i.e. s_{tbl}. Synthesis results give that the area of the register file is roughly 45 μm^2 per bit.

Lastly, the synthesis results for the scheduler subcomponent and remaining register files are shown in Fig. 3.10. The first thing to note is that, in contrast to the previously presented modules, there is a considerable constant part that does not change with the number of ports (the HPU and flit control). The results also show that even the NI scheduler, being one of the more complex parts of the interconnect, achieves clock frequencies well above 600 MHz for NIs with up to eight ports.

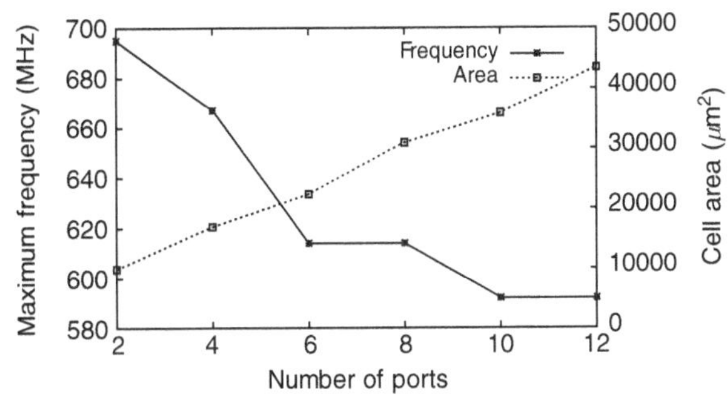

Fig. 3.10 NI scheduler area and frequency

3.5.3 Limitations

Our connection model does not easily extend to (router-based) multicast, as proposed in [157]. The reason is the difficulties associated with implementing a robust flow-control scheme for multiple destinations, i.e. the lack of a scalable end-to-end flow-control scheme for that type of connection. In [157] this problem is avoided by limiting the scope to applications where sufficient space can be guaranteed at design time.

The NI architecture, as presented here, has a design-time fixed slot-table size, and even using a subset of the slots is not possible. The reason is the challenges involved in changing the value globally, across the entire network, without causing inconsistencies between NIs (both during and after the change).

A more severe limitation than the fixed slot-table size is that the (maximum) number of connections, i.e. streaming ports are determined at design time. Moreover, also the buffer sizes are design-time parameters and no adaptations are possible after the network hardware is instantiated. Using SRAMs instead of dedicated FIFOs allows for redistribution within that NI, but one read and write port is needed for every IP (and one for the network) to allow them to access the buffers concurrently. Moreover, having a shared SRAM complicates partial reconfiguration, as we then have to address also fragmentation of the memory space.

In addition to the limitations of the NI architecture, there are four major limitations with the TDM-based arbitration it uses. First, the granularity of throughput allocation is determined by the slot table size. As we shall see this is not a major problem since capacity in the network is abundant and larger slot tables come at a very low cost. Second, latency and throughput are inversely proportional. That is, a channel that requires low latency needs many (properly distributed) slots, potentially wasting network capacity. Third, channels from the same application are temporally composable, even though predictable sharing between them is sufficient. We return to this point in Chapter 4. Fourth, neighbouring routers and NIs rely on *flit-level synchronicity*. We address this issue in Section 3.7 where we show how to implement this assumption in a *mesochronous*, i.e. phase-tolerant [83], network.

3.6 Routers

The responsibility of the routers is to get the data of all connections from one point of the chip to another, based on header contents of the packets. The router, depicted in Fig. 3.11, consists of three pipeline stages, corresponding to a flit size of three phits. The first stage synchronises the input data. Thereafter, a Header Parsing Unit (HPU) determines the output port based on the path encoded in the packet header, similar to what we have already seen in the NI. The selected output port for the specific HPU remains the same until an end-of-packet bit is encountered, similar to what is proposed in [157]. In contrast to the Æthereal architecture [63], the control bits are explicit signals and do not need any decoding, which *removes the HPU from the critical path*. The output port numbers are one-hot encoded before being fed to the multiplexer-based switch which determines the assignment of input to output ports. A phit appears on an output port three cycles after being presented to an input port.

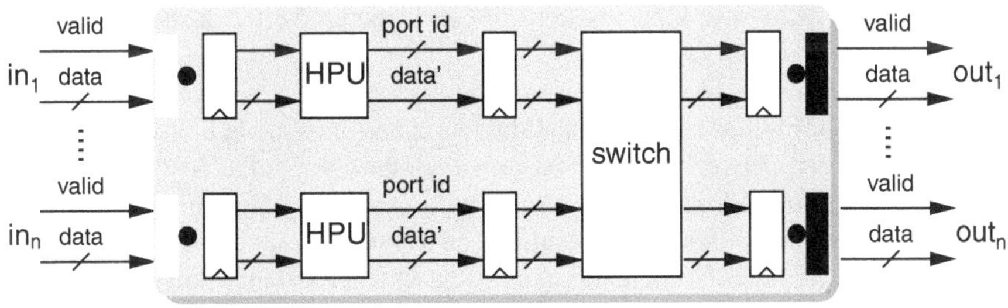

Fig. 3.11 Router architecture

The router has *no routing table and only a one-word buffer per input port.* It also has no link-level flow control and *no arbiter* because contention is avoided through off-line scheduling of the flits. The router also has no notion of TDM slots and blindly forwards the data on the inputs. In contrast to VC-based NoCs [12, 27, 154, 162, 165], the router is not negatively affected by the number of connections or service levels (more and deeper buffers, wider multiplexers), the real-time requirements of the connections (deeper buffers), or the introduction of link pipelining (deeper buffers or flow-control-aware pipeline elements). Pipelining the link can easily be done by moving the input register onto the link itself. This differs from traditional NoCs where pipelined links affect the link-level flow control because the feedback signals take multiple clock cycles [162].

The benefits over the combined guaranteed and best-effort services of the Æthereal architecture are as follows: (1) All connections are isolated and given bounds on their temporal behaviour, thus achieving composability and predictability. (2) Not having best-effort communication *reduces the router to one VC and removes the need for link-level flow control.* This greatly simplifies a mesochronous implementation, thus achieving scalability at the physical level as detailed in Section 3.7. (3) The hardware implementation is much simpler, thus enabling *lower area and higher speed*, i.e. a better cost/performance trade-off, as we shall see in Section 3.6.1.

At design time, the data width of the router (and NI) is determined. Throughout this work we assume 37 bits to match the message formats in Fig. 3.6b. Additionally, each router is dimensioned by choosing the number of input and output ports (potentially different) depending on the network topology. The choice of a physical topology is largely dependent on two things: the logical topology of the connections and physical-level enablers. First, a realisation of the logical topology, i.e. the user requirements, relies on the resource allocation, as discussed later in Chapter 4. The physical network topology must hence be dimensioned such that an allocation can be found. In contrast to works based on virtual channels [12, 27, 154, 162, 165], it is only *the throughput and latency requirements*, and not *the number of connections*, that put requirements on the network topology. The reason is that the router network is stateless [186] and does not limit the number of connections using a link or passing through a router. This is also discussed in Section 2.2.

Second, and equally if not more important are the physical-level enablers, i.e. the routability of high-arity routers and the maximum inter-node link length [162]. With the proposed network, in contrast to many other NoCs, the designer is given full freedom in dimensioning the network, with *any router topology*, *any number of links between two routers*, and potentially an asymmetrical number of inputs and output ports in a router. No distinction is made whether a router port is connected to another router or an NI. Thereby, a router can be connected to an arbitrarily large number of NIs to form either an indirect network (no NI) or what other works have referred to as *concentrated topologies* (more than one NI).

3.6.1 Experimental Results

To determine the area, a number of router instances are synthesised. Figure 3.12 shows the trade-off between target frequency and total area for an arity-5 router with 32-bit data width (used for comparison with [63]). As seen in the figure, the router occupies less than 0.015 mm^2 for frequencies up to 650 MHz, and saturates at 825 MHz. The area and frequency of the router is independent of the number of connections passing through the router, unlike VC-based NoC architectures such as [12, 27, 165], and the synthesis results differ significantly from the combined guaranteed and best-effort service router of Æthereal, that occupies 0.13-mm^2 and runs at 500 MHz, when synthesised in a 130 nm CMOS technology [63]. Just like the conclusion drawn for Mango [27], we see that *the guaranteed-service-only architecture is using less area than a simple fair arbiter*, and comparing our proposed router to that of Æthereal, the difference is roughly 5× less area and 1.5× the frequency when the two implementations are synthesised in the same 90-nm technology.

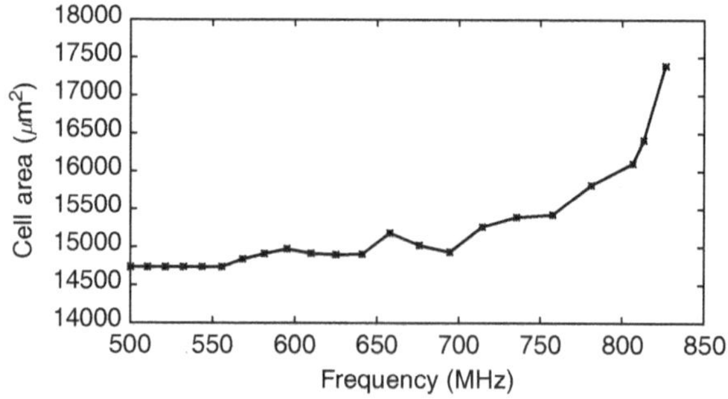

Fig. 3.12 Router frequency and area trade-off

Figure 3.13a shows how the router scales with the arity when synthesised for maximum frequency. We observe a similar drop in maximum frequency as reported in [162]. The stepwise reduction in frequency is caused by the priority-encoded multiplexer tree inside the switch. For every power of two, another level of multiplexers is added. Note that a lower utilisation is to be expected for high-arity routers during placement and routing thus increasing their total area further [162].

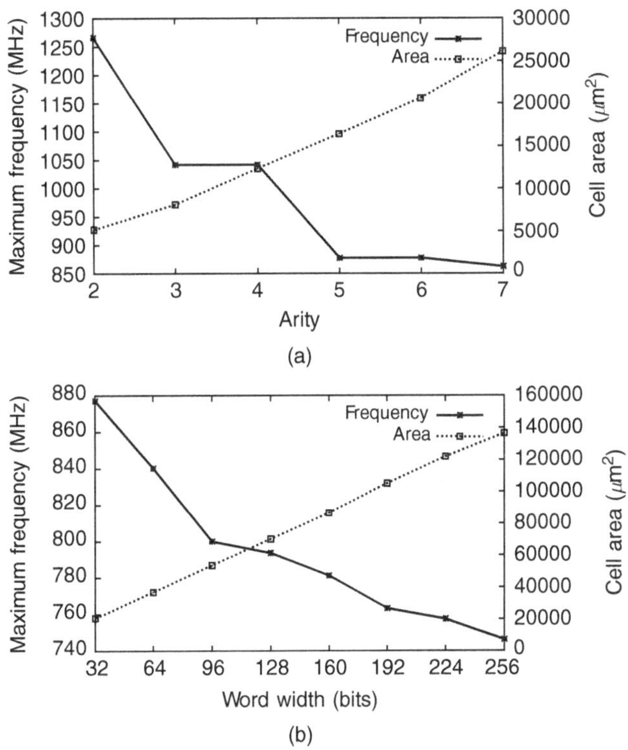

Fig. 3.13 Synchronous router area and frequency for varying arity and data width. (**a**) Varying router arity for 32-bit data width. (**b**) Varying data width for arity-6 router

Figure 3.13b shows how the data width affects the area and obtainable frequency. We observe that the area grows linearly with the word width while the operating frequency is reduced, also with a linear trend. It is clear from our experiments that the router scales to both high arities and wide data widths, thus offering massive amounts of throughput at a low cost, e.g. an arity-6 router offers 64 Gbps at 0.03 mm^2 for a 64-bit data path.

3.6.2 Limitations

In contrast to networks with only best-effort services, resources must be reserved in advance and are not available to other connections. It is important to note, however, that (1) the routers are much cheaper than in the original Æthereal architecture, and (2) reservations do not have to correspond to the worst-case behaviour if this is not needed by the application. That is, an application with no or soft real-time requirements may request resources corresponding to its average-case behaviour.

A limitation that the basic router shares with Æthereal is that the network requires a (logically) *globally synchronous clock*. This places strict requirements on the placement of routers and the distribution of a clock. The link delay problem can be mitigated by using pipelined (synchronous) links [162, 184], thus shortening the wire length. With the proposed router architecture, this is possible by moving the input register, as shown in Fig. 3.11, onto the link itself. However, the clock skew between neighbours must still be sufficiently low to avoid sampling in critical timing regions, severely limiting scalability. This problem is mitigated or completely removed with the introduction of *mesochronous links*.

3.7 Mesochronous Links

The choice of a link greatly affects how sensitive the network is to wire delays and what clock distribution scheme is possible. When the routers and NIs share a clock and thus have the same nominal frequency but different phase relationships, mesochronous links mitigate skew constraints in the clock tree synthesis. Most importantly, mesochronous networks are scalable [29], since the phase difference between regions is allowed to be arbitrary.

Normally, link-level flow control and multiple VCs complicate the implementation of mesochronous (and asynchronous) NoCs [12, 113, 154, 165], due to the increased number of control signals required. The latency involved in the handshakes is also reported to limit the maximum achievable operating frequency [29] (and limiting the throughput in an asynchronous NoC [27]). In our network architecture, there is *no need for link-level flow control and only one VC is required, independent of the number of connections/service levels*. This is a major difference with existing mesochronous/asynchronous networks. The challenge is to provide composable and predictable services without global (phase) synchronicity.

To hide the differences in clock phases we do not only put *bi-synchronous FIFOs between neighbouring elements* [113, 121, 154], but also *allocate a time slot for the link traversal*, thus hiding the difference in phase. The architecture of the link consists of a bi-synchronous FIFO [124, 196] and a Finite State Machine (FSM), as shown in Fig. 3.14. The FIFO adjusts for the differences in phase between the writing and reading clock, where the former is *sent along with the input data* (often referred to as source synchronous), thus experiencing roughly the same signal propagation delay [29]. The FSM tracks the receiver's position within the current flit (0, 1 and 2). If the FIFO contains at least one word (valid is high) the cycle a new flit cycle begins (state 0), the FSM keeps the valid signal to the router and the accept signal to the FIFO high during the succeeding flit cycle (3 clock cycles). Like [154], we assume that the skew between the reading and writing clock is at most half a clock cycle, and that the bi-synchronous FIFO has a forwarding delay less than the number of phits in a flit (1 or 2 cycles) and a nominal rate of one word per cycle. Under these assumptions, the re-alignment of incoming flits to the reading clock ensures that (1) flits are presented to the router in their assigned time slot, i.e. not too early and not too late, and that (2) the three phits of a flit are forwarded to the router in consecutive cycles. The FSM thus re-aligns incoming flits to the *flit cycles of the reading clock*, achieving *flit synchronicity* over a mesochronous link.

With the aforementioned behaviour, the FSM ensures that *it always takes three cycles* (in the reading clock domain) for a flit to traverse a link. This aligns the flit to flit-cycle boundaries by introducing additional latency. The three cycles are enough to absorb the latency of the FIFO and the skew between the writing and reading clock. Moreover, as the phase difference is guaranteed to be bounded, the FIFO is chosen with sufficient storage capacity to never be full (four words). The FIFO does not need to generate a full/accept signal, and *all handshakes are local*, i.e. they stay within the link (and do not limit the clock frequency). There is no need for any link-level flow control or handshakes between the links, routers or NIs.

The mesochronous links are the last part of our network. The links are inserted as part of the dimensioning, i.e. topology selection, based on the envisioned floorplan and physical distances between routers and NIs. In Fig. 3.15, we see the network topology as dimensioned for the example system. As seen in the figure, different links can have a different number of pipeline stages. Note that the NIs have no streaming ports at this stage. During the allocation in Chapter 4, the shells (and IP ports) are mapped to NIs and streaming ports are added.

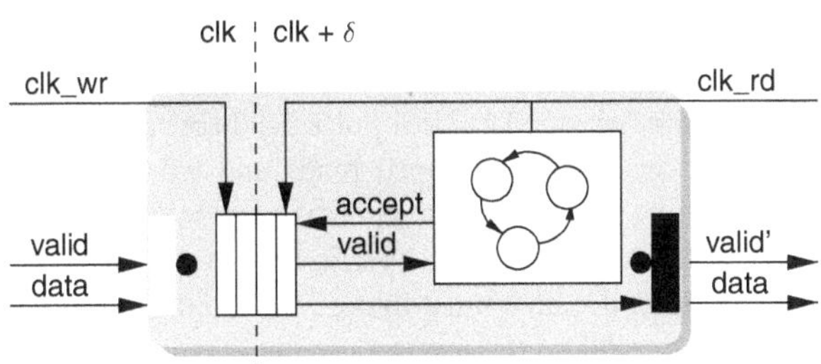

Fig. 3.14 Link architecture

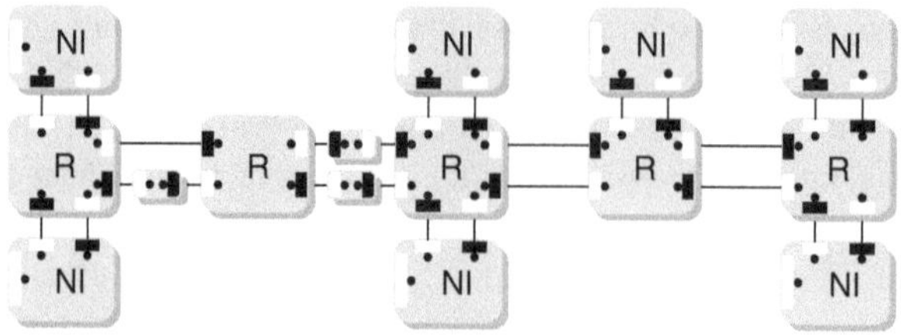

Fig. 3.15 Network topology of the example system

3.7.1 Experimental Results

For the mesochronous links the area is in the order of 1,500 μm^2 when using the custom FIFOs from [196], or roughly 3,300 μm^2 with the non-custom FIFOs from [154]. For an arity-5 router with mesochronous links the complete router with links is in the order of 0.032 mm^2. This is significantly smaller than the mesochronous router in [154], or the asynchronous router in [12], that occupies 0.082 mm^2 and 0.12 mm^2 (scaled from 130 nm), respectively. Also note that these two NoCs offer only two service levels and no composability.

3.7.2 Limitations

The mesochronous links are only applicable if the entire network has the same nominal clock rate. If the routers and NIs are *plesiochronous* (or even *heterochronous*) [121], then adapting the link is not sufficient. Some routers and NIs will be faster than others, and they must be slowed down to guarantee that *input and output is flit-synchronous relative to neighbouring network elements*. In this work we only allow a mesochronous network, and refer the reader to [78] for details on how to address the aforementioned issues.

3.8 Control Infrastructure

With the introduction of the different blocks of the interconnect, we have seen a number of programmable blocks, i.e. the NIs and the buses. In Fig. 2.1, for example, there are nine memory-mapped control ports used for run-time reconfiguration of the interconnect. Using memory-mapped reads and writes, one or more hosts, typically general-purpose CPUs, use those control ports to configure the blocks and thereby open and close connections [68, 168]. To open a connection from the ARM to the SRAM, for example, the host must first configure the address decoder of the target bus. Next, the source NI of the request channel (going from the ARM to the SRAM) must be configured with the path and time slots allocated to the channel. Similarly, the source NI of the response channel also requires configuration. Finally, the initiator bus in front of the SRAM is configured with an allocation according to the specific details of the arbiter implementation. As illustrated by this example, the host has to access multiple different control ports, located at different places in the interconnect topology (and eventually also in the chip layout).

To enable the host to reach the control ports that are distributed across the interconnect and configure the buses and NIs (and potentially also IPs), a *control infrastructure* is needed. A dedicated control interconnect [115, 203], traditionally implemented as a control bus, raises the question whether the control interconnect is scalable at the physical and architectural level. If not, the interconnect on the whole is not scalable. In this work, similar to [46, 68], we address the aforementioned issue by unifying the data and control interconnect, thus creating a *virtual control infrastructure*. Doing so, however, raises a number of challenges that must be addressed. First, resources must be allocated to the *control connections*, as discussed in Section 3.8.1. Second, the control ports (unconnected in Fig. 2.1) must be connected by extending the interconnect architecture with *control buses*. This is discussed in depth in Section 3.8.2. With the control connections and control buses in place, it remains for the host to orchestrate the control infrastructure at run time. We return to this topic in Chapter 5.

3.8.1 Unified Control and Data

The control connections enable the host to reuse the interconnect also for memory-mapped control communication, thus providing a scalable control infrastructure. As the last step of the design-time dimensioning, the control connections are added as an application, named *control*, together with the user applications. In addition, a use-case constraint is added, stating that the control application may run together with all other applications. This is seen at the bottom of Fig. 3.1, as the additions to the communication specification given by the user.

Traditionally, the control connections are implemented using best-effort services [46, 68], as shown in Table 3.2. With best-effort connections, no resource reservations have to be made for the control connections that are both seldom used, and have low throughput requirements. However, no upper bound can be given on the execution time of the control operations, e.g. the time to instantiate an application. In the proposed network that offers no best-effort services, both the control connections and the user-specified connections enjoy composable and predictable services. As we shall see in Chapter 5, this enables temporal bounds on the reconfiguration operations (opening and closing of connections). Furthermore, the control application enjoys the same isolation as the user applications [68] and is not affected when other applications are started or stopped. Hence, the host becomes what is referred to as a *trusted network authority* in [100], and a design fault or a hardware fault in one application cannot affect the service given to any of the other applications in the system (including the control application). The drawback compared to using best-effort communication is the need for permanent resource reservations, because the host must always be able to open and close connections. We minimise

Table 3.2 Control infrastructure implementations

	Shared interconnect	Quality of service
Wolkotte et al. [203]	No	Best effort
Marescaux et al. [115]	No	Predictable
Hansson and Goossens [68]	Yes	Best effort
This work	Yes	Composable and predictable

the cost of the control connections by using the concept of channel trees [71], thus allowing them to share slots (and NI buffers) as further discussed in Chapter 4.

The addition of the control application is necessary to reuse the interconnect for both control and data, but we also have to connect the control ports in Fig. 2.1 to the interconnect. Next, we describe how control buses complement the control connections in reaching the control registers from the host using nothing but memory-mapped communication.

3.8.2 Architectural Components

In addition to the control connections (from NI to NI), control buses are needed to connect all control ports to the network. The control buses address two important issues. First, the host must be able open and close control connections and use them to program the remote NIs and buses. Moreover, it must be able to do so without using the network, to avoid a boot-strapping problem. The host must also be able to be involved in other applications than the control application, e.g. the initialisation and status application in our example system. Second, as already mentioned, it is not only the NIs that need configuration, but also the initiator and target buses. These buses may be in different clock domains than the network and control bus, which must be addressed. We now show how to reuse the modules that have been introduced in this chapter to implement these requirements, through the local and remote control buses illustrated in Fig. 3.16.

3.8.2.1 Local Control Bus

Starting with the local control bus in Fig. 3.16a, we see that the proposed architecture consist of three major building blocks. First, a control bus, i.e. a target bus with a fixed address decoder (Section 3.1), to which the host is directly connected. Second, a number of target and initiator shells (Section 3.3). Third, clock domain crossings (Section 3.4), here only shown as dotted lines. From left to right, according to the numbering in Fig. 3.16a, the local control bus enables the host to:

1. execute boot code,
2. program the local NI,
3. reach remote control buses,
4. program the host's own target bus,

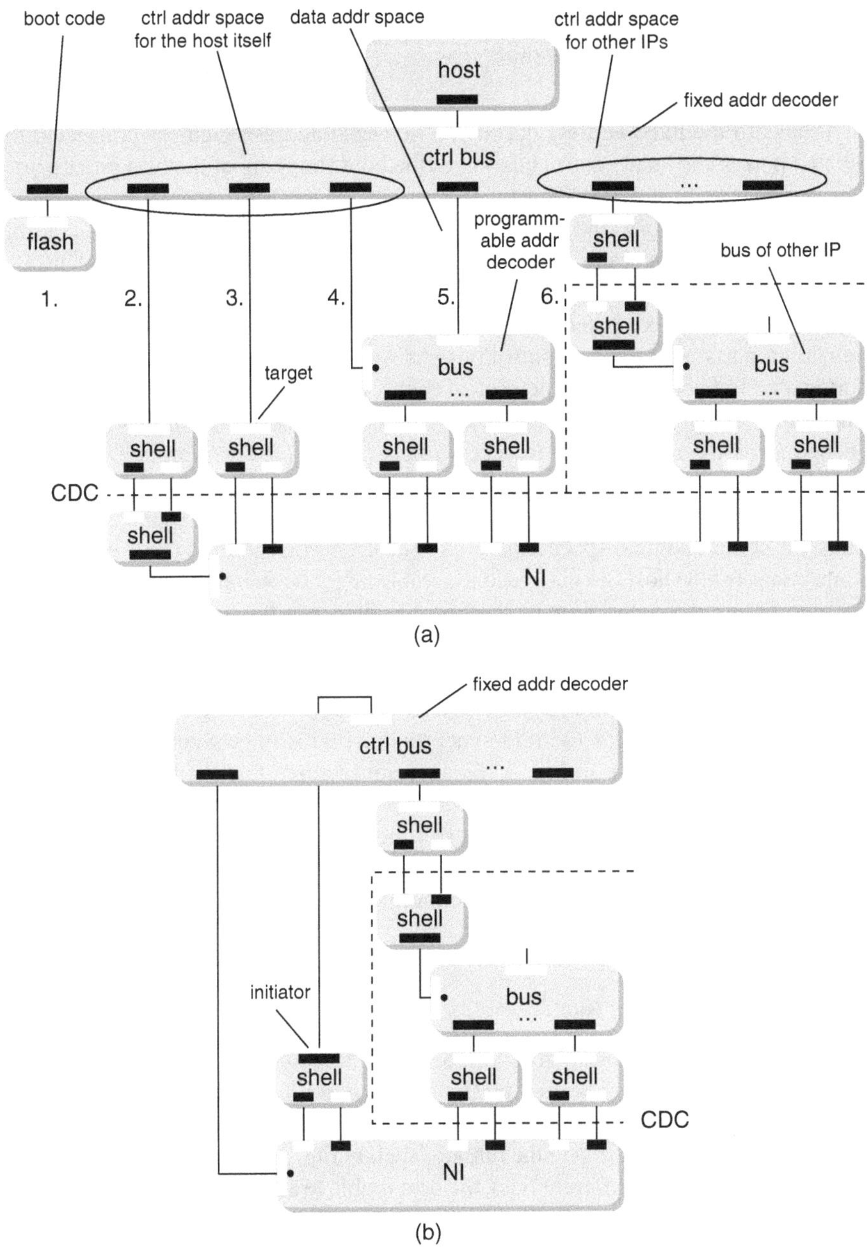

Fig. 3.16 Local (**a**) and remote (**b**) control buses

5. use the aforementioned target bus, and
6. program buses belonging to other IPs.

Thanks to the fixed address decoder, a non-volatile flash memory connected to the first port of the local control bus solves the boot strapping of the host processor.[4] The flash memory holds sufficient information for the host to open one connection to another memory, and continue the execution and instantiation of the control application from there. Furthermore, as only the host has access to the control infrastructure, the boot sequence is secure [47], e.g. for digital rights management.

Once the host has booted, it is able to open and close connections using the remaining ports on the local control bus. As we shall see in Chapter 5, the second port is used to open and close the control connections to other remote control buses. The third port enables the host to use the control connections (once they have been opened using the second port) to access *remote control buses*. Through the fourth port the host programs its own target bus, which is used for other applications than control. Once the bus is programmed, the fifth port can be used for user applications. Thus, the control address space is part of the host's normal address map, and all user connections (of the host) go via a reconfigurable target bus (similar to other IPs). The sixth port is used to configure a bus of an IP other than the host itself. Potentially there are more such ports on the local control bus. A final point to note in Fig. 3.16a is that the control ports belonging to other IPs are potentially in different clock domains, and that the crossings are implemented by connecting a target shell and initiator shell back to back (with bi-synchronous FIFOs in between). Moreover, in contrast to what is shown in Fig. 3.16a, it is not required that the target bus of the host (for user connections) and the buses of other IPs are connected to the same NI as the local control bus. It could, in fact, be five different NIs (only the second and third port of the local control bus must be connected to the same NI). It is also possible to move the fourth port, used to program the address decoder of the host's target bus, to a remote control bus.

3.8.2.2 Remote Control Bus

A remote control bus, as shown in Fig. 3.16b, is accessed via the network and has only a subset of the ports needed by the local control bus. As already mentioned, the host accesses the remote control bus through the target shell in Fig. 3.16a by opening a control connection to the initiator shell in Fig. 3.16b. By opening different control connections (to different NIs), the host is able to access all the remote control buses through the same port. Through each remote control bus in turn, the host can configure the NIs and any buses attached to it. Chapter 5 explains in greater depth how the interconnect reconfiguration is done, using the control infrastructure

---

[4] The alternative, although not currently implemented, is to have a reset state of the local NI where at least one connection is already set up from the instruction and data port of the host to a memory where the boot code is stored.

proposed here. In contrast to the local control bus that is in the host clock domain, we place the remote control buses in the network clock domain. Thereby, clock domain crossings are only required for the control ports of the initiator and target buses, and not for the NI itself. Similar to the local control bus, it is not required for the ports of the IPs to be connected to the same NI as the remote control bus.

3.8.3 Limitations

Our discussion on the control infrastructure is focused on systems with a single host. As we shall see in Chapter 5, this is a reasonable assumption for contemporary SoCs, but is not a scalable solution. Having only one host, however, is no inherent requirement of the proposed methodology. It is possible to have *multiple hosts*, each with its own virtual control infrastructure, and distribute the responsibility of the interconnect across the hosts. In fact, as the control infrastructure is based on the concepts of distributed shared memory, it is even possible to allow multiple host to share the control infrastructure by adding an initiator bus before every (remote) control bus. With multiple hosts, we believe it is possible to scale the proposed solution to large SoCs, with 100 of control ports. Moreover, by establishing a *hierarchy* among the hosts, it is possible to synchronise the access to different control registers such that all modifications are consistent. We return to describe these issues in Chapter 5, there in the context of a single host.

3.9 Conclusions

In this chapter we have presented the building blocks of the interconnect and shown how they are dimensioned. We summarise by evaluating the contribution to the different high-level requirements in Chapter 1.

The interconnect offers *scalability* at the physical and architectural level. The physical scalability is achieved by using GALS for all IPs, and mesochronous links inside the network. This alleviates the designer from strict requirements on clock skew and link delay, thus enabling an effective distributed placement of the network components even for larger die sizes. The architectural scalability stems from the ability to add more links, routers, NIs, shells and buses without negatively affecting the services of the existing components.

The interconnect supports *diverse* IPs and applications. No requirements are placed on the IP's use of the network, thus offering application diversity. Furthermore, thanks to the separation of the network, streaming and bus-protocol stacks, streaming as well as distributed shared memory communication is offered, with an established memory-consistency model (release consistency). The interconnect also support multiple concurrent memory-mapped protocols. Currently, we do however require that the same protocol (stack) is used by all the buses in the system.

In the proposed network, just as Æthereal, *composability* is based on contention-free routing, with complete isolation of connections. Composable sharing of memory-mapped targets is provided through the use of target buses and atomisers.

Predictability is provided as the latency and throughput in the network follows directly from the waiting time in the NI (plus the time required to traverse the path), and the fraction of slots reserved, respectively. Performance analysis at the application level is possible by modelling the entire interconnect in a dataflow graph, something we discuss in depth in Chapter 6.

To enable *reconfigurability* we place a virtual control infrastructure on top of the interconnect. This results in a unified data and control interconnect, with reuse of existing shells and buses, where control enjoys the same isolation and temporal bounds as other applications. The reconfigurability is thus scalable, composable, predictable and additionally secure and robust.

The contributions to the *automation* lies primarily in selecting, adding and connecting components, pushing the complexity of allocating and using those resources to future chapters. Nevertheless, the three steps in Fig. 3.1 correspond to three tools in the design flow for bus and shell instantiation, network topology generation, and addition of the control infrastructure. Additionally, the area models described throughout this chapter are included in an area-estimation tool that quickly shows the total area requirements of the interconnect, and the distribution across different components.

Obviously, all the qualitative properties, outlined above, do not come for free. We do believe, however, that the interconnect provides *a good performance normalised to silicon area*. The important observation about the silicon area is that the interconnect is dominated by the NIs that in turn are dominated by their buffers. Buffer sizing is therefore of utmost importance, and we return to see the impact of the NI buffers in Chapter 6. The silicon area of the interconnect, in the context of a complete system, is exemplified in Chapters 7 and 8.

Chapter 4
Allocation

Resource allocation is the process of assigning the resources of the interconnect of Chapter 3 to the individual applications, based on their requirements and the use-case constraints. Allocations are created and verified at *design or compile time*. Hence, *run-time* choices are confined to choosing from the set of fixed allocations. While limiting the run-time choices to a set of predefined use-cases, this is key as it enables us to guarantee, at compile time, that all application constraints are satisfied, and that all the use-cases fit on the given hardware resources. In this chapter, which corresponds to Step 2 in Fig. 1.8, we present an allocation algorithm that matches the application requirements with the interconnect resources.

To realise a logical connection between the IP ports, the resource-allocation part of the design flow, shown in Fig. 4.1, must,

1. in case of distributed memory communication, allocate a target port on the target bus. Similarly, allocate an initiator port on the initiator bus in case of shared memory communication. The selection of a bus port implicitly selects a shell on each side of the network (there is a one-to-one mapping). After this step, every connection is allocated two communicating pairs of streaming ports.
2. if the resource allocation is done at design time, then decide where in the network topology the shells (or IPs in the case of a streaming protocol) are connected. The streaming ports from the previous step are consequently connected to NIs. After this step, every connection is allocated an NI on each side of the network (possibly the same).
3. for the two channels that constitute the connection, select paths through the router network, starting and ending at the NIs determined in the previous step. On the selected paths, allocate an appropriate set of time slots such that the throughput and latency requirements are met. After this step, the allocation is complete and the logical connection can be instantiated.

The first step is only needed in those cases where a connection passes through an initiator or target bus, exemplified in Fig. 4.2 by the decoder's connection from the ARM to the SRAM, illustrated with dotted arrows. In Chapter 3 we only dimensioned the buses, i.e. we only decided the number of bus ports, and did not yet decide what connection uses what port in what use-case. The allocation of ports to

A. Hansson, K. Goossens, *On-Chip Interconnect with Aelite*, Embedded Systems, DOI 10.1007/978-1-4419-6865-4_4, © Springer Science+Business Media, LLC 2011

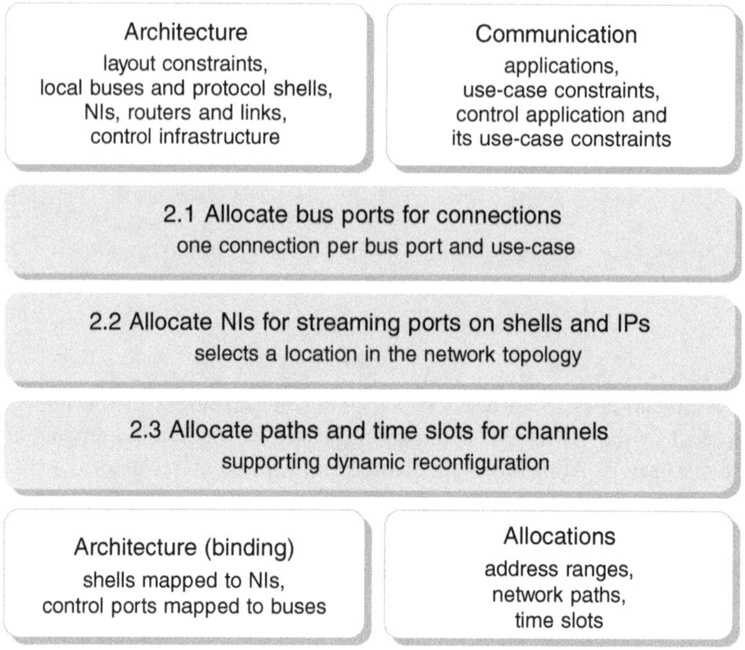

Fig. 4.1 Allocation flow

connections is made by a greedy heuristic that iterates over the applications, based on the number of use-cases they span. The intuition behind this ordering is that the ports used by an application cannot be used by any other application that runs concurrently. Hence, the more use-cases an application spans, the more restrictions are imposed on the not-yet-allocated applications.

After allocating the bus ports on both sides of the network, each connection is associated with two pairs of streaming ports. In case of a memory-mapped protocol, these streaming ports are located on the shells. For a streaming protocol, on the other hand, the ports are those of the IP itself (possibly with a clock domain crossing in between). After performing the bus-port allocation, the succeeding steps of the allocation flow no longer looks at connections between IPs, but rather individual channels between streaming ports. These *mappable ports* are shown in Fig. 4.2. As we shall see, the remaining two steps of the compile-time allocation flow perceive applications as *channels between mappable ports*.

After the bus ports are assigned to connections, the second step decides the spatial mapping of mappable ports to NIs, i.e. where in the *given network topology*, shown in Fig. 3.15, to connect them. The mapping has a large impact on the succeeding steps, and greatly affects the energy, area and performance metrics of the system [86, 148]. More importantly, the mapping influences the ability to satisfy the application guarantees during succeeding steps [70]. Traditionally [85, 86, 133–135], mapping is perceived as a special case of the NP-complete *Quadratic Assignment Problem*

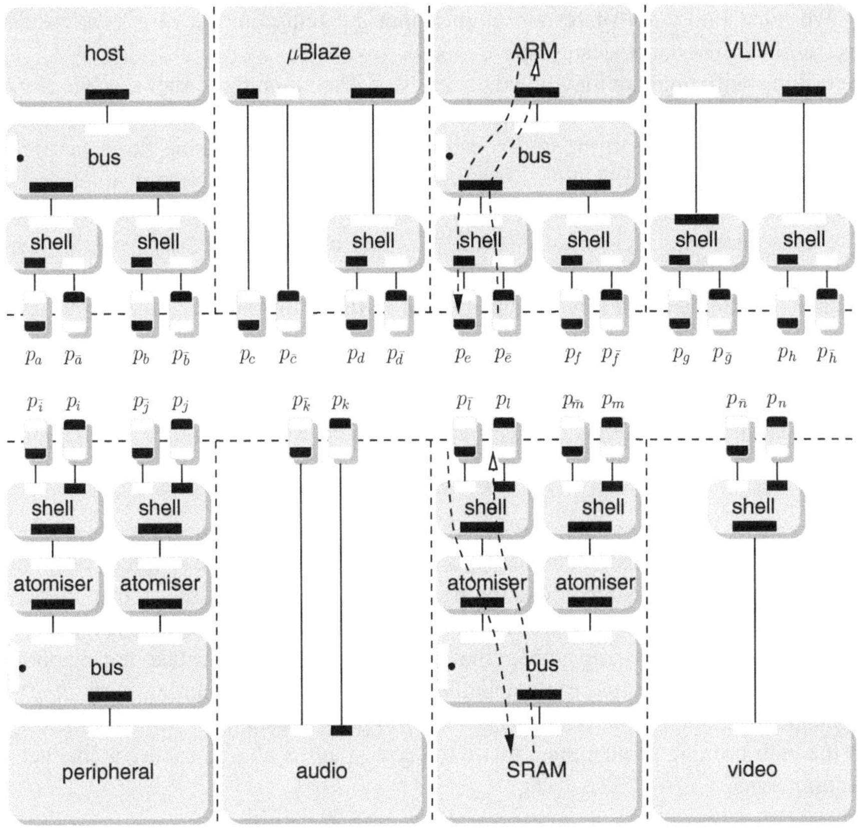

Fig. 4.2 Mappable ports of the example system

(QAP) [155]. Intuitively, the QAP can be described as the problem of assigning a set of mappable ports to a set of locations (NIs) with given distances (path length) between the locations and communication weights between the ports. The goal is then to map the ports onto locations in such a way that the sum of the products of weights and distances is minimal. As we shall see, the introduction of constraints such as latency requirements and slot alignment calls for other approaches than the traditional QAP formulation.

The last step of the allocation flow concerns the use of the network itself, and determines the channel path and slots. The resulting allocation must adhere to the channel's throughput and latency requirements. The allocation algorithm proposed in this chapter takes a network-centric approach and excludes the effects the buses and IPs (and also flow control between these components) have on the end-to-end temporal behaviour. We return to discuss the rationale behind this decision in Chapter 6 where we also demonstate how to incorporate the end points and dimension the NI buffers using dataflow analysis techniques.

We must find a set of resources such that the requirements of every channel are fulfilled. The large design space, coupled with the many constraints, leads to algorithms with high computational complexity. The mapping of shells to NIs alone is NP-hard, and so are the constrained path selection and slot allocation. To simplify the problem, the resource allocation is therefore done using heuristic algorithms. In [62, 85, 86, 133, 134, 136] the mapping and path selection is functionally decomposed into modules on the basis of a flowchart. Each module has its separate constraints and optimisation goals e.g. map by clustering based on throughput requirements and route based on contention and hop count. However, mapping decisions anticipate and rely on the abilities of the path-selection (and in our case slot-allocation) algorithm to find feasible routes between the locations decided by the mapping algorithm. In contrast to the large body of optimising mapping and path-selection algorithms that only look at channel throughput requirements and link capacity [85, 86, 133, 134, 136], we additionally adhere to the latency requirements of the channel, and the restrictions imposed by the pipelined virtual circuits (i.e. slot alignment).

Instead of basing mapping decisions on distance and volume estimations, and path-selection decisions on simple capacity metrics, we propose a tighter coupling between the three resource allocation phases [70]. As we shall see in this chapter, and as illustrated in Fig. 4.3, we exploit the close correspondence between the two spatial allocation problems and let the path-selection algorithm take the mapping decisions by incorporating the mappable ports (shells and streaming ports of IPs) in the path-selection problem formulation. Similarly, we include the slot allocation in the path pruning to guarantee that a traversed path is able to satisfy a channel's requirements.

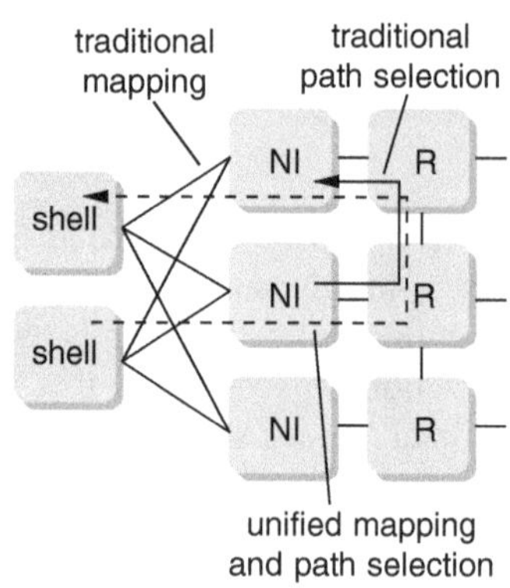

Fig. 4.3 Unified mapping and path selection

Before discussing the actual algorithms that transform the requirements and constraints into allocations, we start by showing how to exploit mutually exclusive channels not only between applications, but also within an application, by sharing time slots (Section 4.1). We proceed by introducing the problem formulation (Section 4.2). Then, the algorithm for allocating network resources (Section 4.3) follows. We end this chapter with experimental results (Section 4.4) and conclusions (Section 4.5).

4.1 Sharing Slots

Contention-free routing is non-work-conserving [205], i.e. the arbiter in the NI may be idle even when there is data available to be sent (on connections other than the one currently scheduled). Although the time in the router network is minimal (no contention), flits have to wait for their slots in the NIs, even if there are no other flits in the network. As a result, the average latency is relatively high even in a lightly loaded system [26, 205]. Moreover, latency and throughput are inversely proportional. That is, the larger the slot table and the fewer slots are reserved, the higher the latency as the distance between reserved slots increases. Another issue with the contention-free routing is that it is a type of frame-based arbitration, where low throughput channels in combination with limited allocation granularity are a source of over-dimensioning [128] and can lead to under-utilisation of resources [171, 205]. The latter is, for example, the case for the control connections that all have minuscule throughput requirements.

The discretisation and latency effects are mitigated through the introduction of *channel trees*, where time slots are reserved for sets of communication channels that form a tree around a shared source or destination port [71]. As illustrated in Fig. 4.4, this formation of trees occurs naturally at memory-mapped target ports shared by multiple initiators and at memory-mapped initiator ports that use distributed memory, i.e. access multiple targets. For an initiator with distributed memory, the request channels diverge, and the response channels converge. Correspondingly, for a shared target, the response channels diverge and the request channels converge. The set-top box SoC shown in Fig. 4.5 illustrates how channels are formed in an actual system instance. In this system, trees are formed around the external SDRAM, but also around the two CPUs, responsible for the peripherals and accelerators belonging to their respective task domain.

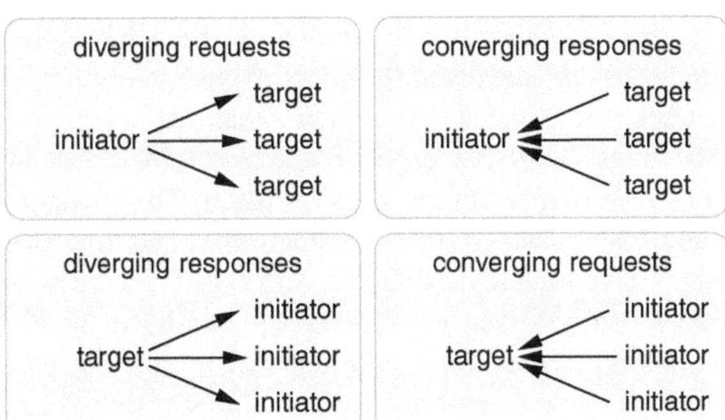

Fig. 4.4 Channel trees

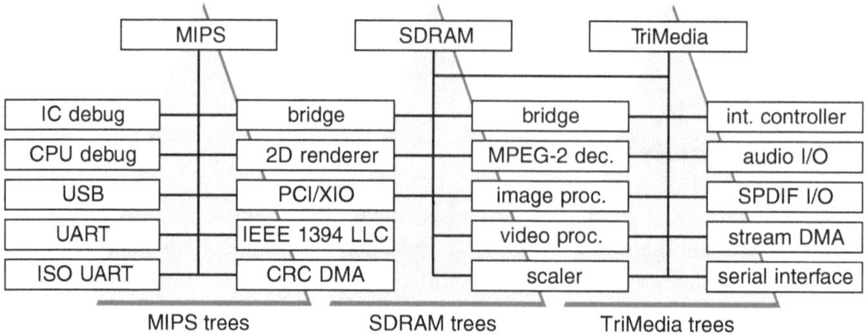

Fig. 4.5 The Viper set-top box SoC [50]

As described in Section 2.1.1, slot allocation is traditionally only correct if every slot of a link is allocated to at most one channel per use-case. However, by reserving slots for trees of channels *within one application*, we are able to allow channels used in a mutually exclusive manner to share a slot reservation. The sharing enables us to either keep the sum of discretised requirements (first discretise, then sum) and reduce the average-case latency, or to reduce the number of slots by reserving the sum of non-discretised requirements (first sum, then discretise). Figure 4.6a illustrates an example in which four channels, c_A, c_B, c_C and c_D require $\frac{1}{40}$, $\frac{2}{40}$, $\frac{3}{40}$ and $\frac{4}{40}$ of the link capacity respectively. The slot reservation is shown both in the form of the slot table of the source NI, and along every hop of the channels' paths. Due to the granularity, the slot table requires at least 40 slots. It is possible to use a smaller slot table at the expense of over-allocation, e.g. use a table with only four slots and assign a single slot to each channel. A minimum of four slots is still required though. In Fig. 4.6b, the slots that were previously reserved for c_A, c_B, c_C and c_D are now collectively reserved for the set that comprises all four channels. Hence, each of the channels may access the network in slots 0–9. In addition, with the introduction of $\{c_A, c_B, c_C, c_D\}$, it is possible to redistribute the ten slots equidistant over the TDM table. This distribution not only minimises the worst-case waiting time for a slot but also enables a reduction of the table size by a factor of ten. The reduced table has four slots and requires only a single slot allotted to the newly introduced channel tree. We return to discuss the details of this example in Section 4.3.3.

The channel trees are isolated from each other and other individual channels. Thereby, the trees cover the entire spectrum from reservations per channel [105, 122, 168, 174] to reservations per port [202]. The channel trees do, however, require an additional level of (intra-application) scheduling to ensure that the constituent channels are mutually exclusive, and that they are scheduled so that real-time guarantees can be given to them [171]. As shown in Fig. 4.4, we distinguish between *diverging* and *converging* channel trees. Request channels form diverging trees at initiator ports, and response channels at shared target ports. Correspondingly, response channels converge at initiator ports and request channels at target ports. For a diverging tree, the contention resolution requires only minor

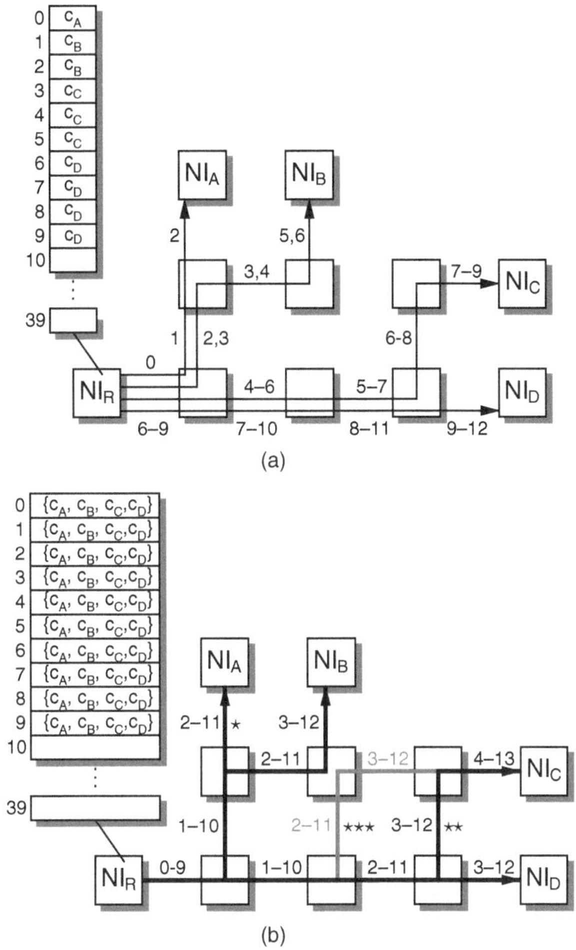

Fig. 4.6 Diverging channels on a 2 × 3 mesh, without (**a**) and with (**b**) trees

adaptations of the source NI [71]. The reason is that they all originate from the same NI and that the contention resolution is *local*. A tree of converging channels, however, requires more elaborate forms of contention resolution since the scheduling decisions are distributed across the source NIs of the channels and they only meet along their path to the destination NI. The problem of contention resolution within the converging tree is thus distributed. In the simplest case, with an initiator at the root of the tree, the mutually exclusive use of the response channels can be guaranteed by only allowing one outstanding request (and thus only one response travelling back to the initiator). This is, however, at the cost of pipelining and parallelism, and in the general case a second arbitration scheme must be added on top of the TDM.

In contrast with [71], where additional arbitration is added to the NI,[1] we restrict ourselves to channel sets where the mutual exclusion is explicitly managed by the IP at the root of the channel tree. By doing so, we push the responsibility for management of the individual channels outside the NI, leaving the NI architecture unaffected and unaware of the sharing of slots. The explicitly managed channel trees fit well with our virtual control infrastructure, where the host sequentially accesses a large number of targets, each with their own low-throughput request and response channels. The channel trees enable the control infrastructure to grow to a large number of NIs without consuming more than one slot per link. Even though we limit the scope to channel trees managed by the IP, the sharing of slots must be considered in the resource allocation, as discussed later.

4.2 Problem Formulation

An essential part of the resource allocation is the specification of application requirements and constraints. The formulation enables the user to express what qualitative and quantitative properties a valid solution should have. It also enables partial specifications, where steps in the allocation are either completely left determined (e.g. shells already mapped), or only partially performed (e.g. some connections already assigned to buses).

The specification can be divided into four parts, related to, respectively, the application requirements, the network topology, the resource allocations and finally the residual resources. An in-depth description of each part follows.

4.2.1 Application Specification

The first part of the problem formulation is the specification of what requirements the applications have on the interconnect. Throughout this chapter we use capital letters to denote sets.

Definition 1 From the perspective of the network, an *application* a is a directed non-multigraph, $a(P_a, C_a)$, where the vertices P_a represent the *mappable ports*, and the arcs C_a represent the set of *channels* between the ports. There is never more than one channel between a given pair of (mappable) ports. Each channel in the application $c \in C_a$ is associated with a *minimum throughput constraint*, $\rho(c)$, and a *maximum latency constraint*, $\theta(c)$. The source and destination of c are denoted $src(c)$ and $dst(c)$.

The end points of the application channels, i.e. the mappable ports, are ports that connect to NIs, as illustrated in Fig. 4.2. For IPs with streaming ports, the mappable

[1] These architectural additions are at this moment only available in SystemC and not in the synthesisable HDL implementation of the proposed interconnect.

ports correspond directly to the IP ports, potentially connected via a clock domain crossing. For memory-mapped ports, however, the application formulation excludes the buses and shells, and takes a network-centric approach. The reason we exclude the buses and shells is that their (temporal) behaviour is largely dependent on the specific protocol and IP behaviour (in the case of an initiator bus). We exemplify how to include specific instances of the buses and shells in the end-to-end performance analysis in Chapter 7.

The filter application, introduced in Fig. 1.5a and here shown in Table 4.1, serves to illustrate how the compile-time flow perceives an application. The filter has four channels, interconnecting eight ports, two on the audio codec (p_k and $p_{\bar{k}}$), two directly on the μBlaze (p_c and $p_{\bar{c}}$), two on the shell of the μBlaze (p_d and $p_{\bar{d}}$) and two on one of the shells of the bus in front of the SRAM (p_l and $p_{\bar{l}}$, decided by the bus-port allocation). The names of the ports correspond to the notation in Fig. 4.2. The two channels between the processor and the audio codec carry only raw audio (streaming) data, and the two channels corresponding to the communication between the processor and SRAM carry (memory-mapped) request and response messages, respectively. Thus, the throughput requirements of the latter reflect not only the data seen by the ARM, but also message headers added by the shells (depending on the protocol and data width). The applications are refined into a specification according to Definition 1 after Step 2.1 in Fig. 4.1.

Table 4.1 Specification of the *filter* application

Source	Destination	Throughput (Mbps)	Latency (ns)
p_c	$p_{\bar{k}}$	1.5 Mbps	1,000 ns
p_k	$p_{\bar{c}}$	1.5 Mbps	1,000 ns
p_d	$p_{\bar{l}}$	5 Mbps	500 ns
p_l	$p_{\bar{d}}$	3 Mbps	500 ns

Irrespective of what data a channel carries, the requirements on temporal behaviour are expressed by two values, similar to the well-known framework of *latency-rate servers* [185]. The throughput requirement is a *lower bound* on the average throughput from the input queue in the source NI to the output queue in the destination NI. The latency requirement is an *upper bound* on the time a single data item (word) can spend from being at the head of the input queue in the source NI, until it is placed in the output queue of the destination NI. It should be noted that the bounds on latency and throughput do not include the effects of end-to-end flow control. In other words, the network resource allocation assumes that the scheduler in the source NI is only governed by the slot table and never stalled due to lack of credits. We return to discuss how to take end-to-end flow control into account in Chapter 6.

The throughput and latency requirements are inputs to the compile-time flow. It is therefore up to the user to decide how the numbers are derived, e.g. by back-of-the-envelope calculations, guesstimates based on previous products, high-level simulation, code analysis or analytical models. It should be noted that the design flow does not require the user specification to capture the worst-case behaviour. If a worst-case allocation is not needed, for example in the case of a soft real-time application, then the communication requirements may reflect the expected or average-case use of the interconnect (or any other value decided by the user). Note, however, that the composability and predictability of the interconnect are not affected by over- or under-provisioning of resources. In the case of over-provisioning, the result is an unnecessarily costly interconnect. In the case of under-provisioning (of throughput or buffers), the interconnect slows the communication down to be means of flow control at the edges (NIs). The formulation of applications and requirements used in this work are also compatible with existing specification formats [62], which we return to discuss in Chapter 6.

As discussed in Section 4.1, mutually exclusive channels are allowed to form slot-sharing trees [71].

Definition 2 The *channel trees* correspond to a partition of C_a into *jointly exhaustive* and *mutually exclusive* sets. The equivalence relation this partition corresponds to considers two elements in C_a to be equal if they have the same source or destination port. The channel set, i.e. equivalence class, of a channel c is denoted $[c]$.

For the resource allocation, the channel trees add additional constraints, as the channels constituting a tree share their time slots. For simplicity, the allocation algorithms currently assume that the throughput and latency requirements for all channels in a tree are the same.

Definition 3 The *set of applications* is denoted A. We define the *complete set of ports* P as the union over all applications, $P = \bigcup_{a \in A} P_a$. Similarly, the *complete set of channels* $C = \bigcup_{a \in A} C_a$.

In our example system, $A = \{$*filter*, *player*, *decoder*, *game*, *init*, *status*, *control*$\}$. Note that the resource allocation perceives the control application (added as part of the control in infrastructure in Section 3.8.1) like any other application. The set of ports P contains all the ports that are to be mapped to NIs. We have already seen 28 of those ports in Fig. 4.2, with the port pairs p_a through p_n, but in fact our example system has 42 mappable ports in total. The remaining 14 ports belong to the control infrastructure, with one port pair per NI, here after denoted p_0 through p_6, for the seven NIs respectively.

Definition 4 A *use-case* $u \subseteq A$ is a set of applications that run simultaneously. Let U denote the *set of use-cases*. Let $U_a \subseteq U$ denote the *set of use-cases containing an application* a, and similarly let $U_c \subseteq U$ denote the *set of use-cases containing a channel* c.

A number of example use-cases, e.g. $u = \{$*status*, *decoder*, *player*, *control*$\}$, are already shown in Fig. 1.6b (after the addition of the control infrastructure). An undirected graph, exemplified by Fig. 1.6, expresses which applications may run in parallel. With this constraint formulation, every clique in the graph corresponds to a use-case. Hence, the process of going from the user-specified constraint graph to a set of use-cases is merely a question of applying a clique-detection algorithm [151]. Doing so for our example system, we get six possible use-cases listed in Table 4.2.

Table 4.2 Example system use-cases

Use-case	Applications
u_0	{*control*, *init*, *filter*}
u_1	{*control*, *init*, *player*}
u_2	{*control*, *status*, *decoder*, *filter*}
u_3	{*control*, *status*, *decoder*, *player*}
u_4	{*control*, *status*, *game*, *filter*}
u_5	{*control*, *status*, *game*, *player*}

With the exception of the control application, we have already seen these six use-cases in Fig. 1.6b. Note that Definition 4 restricts the set of use-cases, i.e. application combinations, but does not restrict the order in which the use-cases appear. It is thus possible to go between any pair of use-cases.

To be able to constrain the mapping of ports to NIs, we allow port groups and map groups instead of individual ports.

Definition 5 The *set of mapping groups* correspond to a partition Q of P, where the elements of Q are *jointly exhaustive* and *mutually exclusive*. The equivalence relation this partition corresponds to considers two elements in P to be equal if they must be mapped to the same location, i.e. the same NI. The equivalence class q of a port p is denoted $[p]$.

The introduction of mapping groups is, in fact, necessary to ensure that the two ports in a streaming port pair, for example p_a and $p_{\bar{a}}$ in Fig. 4.2, are both mapped to the same NI. As already discussed in Chapter 3 this is necessary due to the bidirectional nature of a connection (data travelling on one channel relies on the return channel for the delivery of flow-control credits). Additionally, the grouping of ports enables the designer to express more complex physical layout requirements, similar to what is proposed in [162]. It is, for example, possible to group ports from one or more IPs depending on voltage and frequency islands in the layout. For an example of how mapping groups are used, consider the μBlaze in Fig. 4.2. The streaming port pair of the processor forms one group, just as the two streaming ports on the processor's shell. To ensure that all four ports (p_c, $p_{\bar{c}}$, p_d and $p_{\bar{d}}$) are mapped to the same NI, we add a port group to the architecture description, as illustrated by the constraints in Appendix A.

4.2.2 Network Topology Specification

The second part of the problem formulation, and also part of the input specification, is the network topology, given by the user.

Definition 6 The network topology is a strongly connected directed multigraph. The set of vertices N is composed of three mutually exclusive subsets, N_r, N_n and N_q. N_r contains *routers and links*, N_n *network interfaces*, and N_q *mapping-group nodes*. The latter are dummy nodes to allow unmapped mapping groups to be integrated in the network graph. The number of mapping-group nodes is equal to the number of port groups to be mapped, $|N_q| = |Q|$. The NIs and mapping-group nodes together form the set of *mappable nodes* $N_m = N_n \cup N_q$.

The set of edges L contains the physical network links between nodes in N_r and N_n, and dummy mapping links that connect nodes in N_q to nodes in N_n. A router is allowed to have any number of links while NIs and link pipeline stages always have one egress link and one ingress link. Source and destination of l are denoted $src(l)$ and $dst(l)$.

Figure 4.7 shows the network model for our example system in Fig. 2.1 (before the mapping to NIs is completely decided upon). We clearly see that the core of the graph is merely reflecting the physical network topology. Routers and link pipeline stages, i.e. the nodes in N_r, are grouped together since there is no need to distinguish between the two. From a resource-allocation point of view, there is no difference between a 1-arity router and a link pipeline stage. Similar to the routers and link pipeline stages, there is a direct correspondence between the NIs in the architecture and the nodes in N_n. The naming of ingress and egress links, as shown in Fig. 4.7, refers to the direction relative to the NIs. In contrast to the nodes in N_r and N_n,

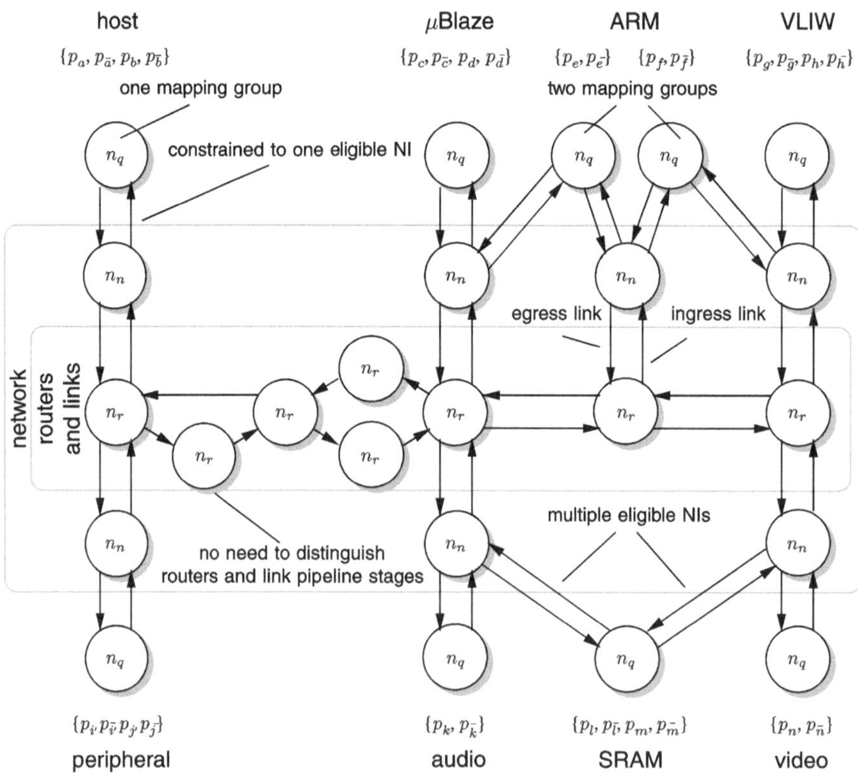

Fig. 4.7 Network model of the example system

the mapping-group nodes N_q do not correspond to any physical network block, but merely act as the starting location for any port group not yet mapped to an NI. As we shall see, N_m contains all nodes to which the elements of Q can be mapped.

The mapping constraints introduced in the application specification are used to ensure that a group of ports are mapped together. We also enable the designer to restrict to which NIs the group is mapped.

Definition 7 The set of *eligible NIs* $\epsilon(q) \subseteq N_n$ constrains the mapping of a group q to a subset of the available NIs.

Rather than allowing the design flow to choose any NI as the location for a mapping group, it is thus possible to constrain the choice *to a set of NIs* that are, e.g., on the perimeter of the chip layout (for an off-chip memory controller), or close to each other physically (for a large IP with multiple ports). Note, however, that even with multiple eligible NIs, all ports within a group are mapped to a single NI. In Fig. 4.7 we have chosen to let all IPs but the ARM have their ports in a single mapping group. As a result, there are nine mapping groups and consequently nine mapping-group nodes in the graph. The topology of links between the mapping-group nodes and the NIs is determined by the eligible NIs. To automatically arrive at a mapping of ports to NIs similar to what is depicted in Fig. 2.1, we have constrained the mapping by having only one eligible NI for all groups besides the ports of the ARM and the SRAM. For a typical system, we start with far less restrictive constraints than what is shown in Fig. 4.7, and even have mapping groups where *all the NIs* in the system are considered eligible, i.e. $\epsilon(q) = N_n$, for all the groups. For clarity, we refrain from cluttering the figure with such an example and also exclude the mapping-group nodes of the control infrastructure. In addition to what is shown in Fig. 4.7, each control bus constitutes one mapping group, and is constrained to a single NI. The specification of eligible NIs is exemplified in Appendix A.

4.2.3 Allocation Specification

The application requirements and network topology are both inputs to the compile-time flow and together constitute all we need to formulate *the allocation problem*, captured by a mapping function and an allocation function.

Definition 8 The mapping of port groups onto the mappable nodes of the network is determined by the *mapping function* $map_i : Q \rightarrow N_m$ that maps port groups to mappable nodes. As the resource allocation progresses, this function is refined. Our starting point is an initial mapping, map_0, where every $q \in Q$ is mapped to a unique $n_q \in N_q$. Let the set of *mapped ports* P'_i denote those elements $p \in P$ where $map_i([p]) \in N_n$. From our definition of map_0 it follows that $P'_0 = \emptyset$.

The range of map_0 initially covers only N_q, i.e. all port groups are mapped to the dummy mapping nodes. Note that it is possible for a port group q to have a set of eligible NIs containing only one item, $|\epsilon(q)| = 1$, either due to user constraints, or because the allocation algorithm is executed with a fixed hardware architecture where the mapping is already decided. We have already seen multiple examples of such constraints in Fig. 4.7. Even in those cases, the allocation algorithm starts by mapping the group to a mapping-group node, but the latter is then only connected to a single NI. As the algorithm progresses, the range of map_i covers both N_q and N_n partially. Thus, some groups are mapped to NIs while others are still mapped to mapping-group nodes. Successive refinements of map_i progressively replace elements of N_q with elements of N_n until a final mapping is derived, where the range of map_i exclusively contains elements of N_n. At this point, all ports are mapped to NIs, i.e. $P'_i = P$.

In addition to the mapping of source and destination port to NIs, each channel requires a path between these two NIs, and a set of time slots on the links of the path.

Definition 9 An *allocation function* $alc_i : C \to \text{seq}\, L \times T$ associates each channel with a path ϕ and a set of time slots $T \in \mathcal{P}(S_{\text{tbl}})$, for a given iteration. A path $\phi \in \text{seq}\, L$ is a sequence of links. The same slot table size s_{tbl} is used throughout the entire network and S_{tbl} denotes the set of natural numbers smaller than s_{tbl} (including zero). $\mathcal{P}(S_{\text{tbl}})$ is the powerset of S_{tbl}, i.e. the set of all subsets of S_{tbl}. The time slots are given relative to the first link, head ϕ, as a set of natural numbers. Initially, for alc_0, all channels have an empty path and no time slots are reserved.

Definition 9 clearly shows that each channel is associated with a single set of resources. That is, the path and slots of a channel (and thus also the connection and application it is part of) are always the same, irrespective of other channels.[2] As we have seen in Section 2.6, having an allocation that is independent of the current use-case is key in enabling partial run-time reconfiguration on the granularity of applications.

As previously discussed in Chapter 3, the slots are shifted one step for every hop along the path. Hence, it suffices to specify the slot allocation relative to the first link in the path. The reason we choose the first link is that, for the final path (without any links to mapping-group nodes), the first link corresponds to the ingress link of the source NI, i.e. the NI where the slot allocation of the channel in question is enforced. We elaborate on this when discussing how channels are opened in Chapter 5.

4.2.4 Residual Resource Specification

The final part of the problem formulation is neither part of the input nor the output of the resource allocation. During the allocation of channels, however, we must know what resources are already occupied and what resources that are still available for the not-yet-allocated channels.

---

[2] Even channels in trees have their own path and set of slots. The allocation algorithm, however, ensures that they all have the same slots.

Definition 10 A *slot table* $t \in \text{seq}\, C_a$ is a sequence of length s_{tbl} where every element is a set of channels (in a channel tree). An empty slot is represented by $\varnothing$. The set of *available slots* for a channel c in a slot table t is denoted $\sigma(c, t) \in \mathcal{P}(S_{\text{tbl}})$.

In contrast to the allocation that contains a set of slots per channel, a slot table (one per link), takes a network-centric perspective and conveys for every time slot what channels are (potentially) using a link. This is exemplified in Fig. 4.8a, where four different slot tables are shown after the allocation of channels c_0 and c_1, with c_0 and c_3 being in the same channel tree. Due to the channel trees, many channels may occupy the same time slot (c_0 and c_3), although of course not at the same time. The available slots include both empty slots and those slots where a channel is already part of the reservation. For example, channel c_3 has two available slots on the western ingress link, $\sigma(c_3, \langle\{c_0, c_3\}, \{c_0, c_3\}, \{c_1\}\rangle) = \{0, 1\}$, despite there being no empty slots on the link in question.

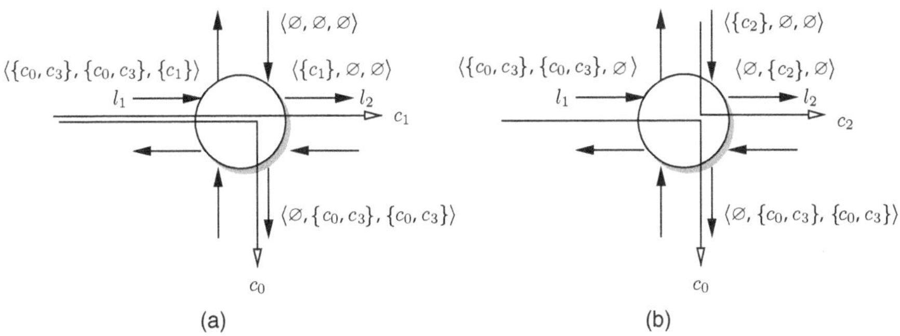

Fig. 4.8 Residual resources for use-case u_0 (a) and u_1 (b)

As Fig. 4.8 already shows, each combination of applications gives rise to its own set of residual resources. For example, channel c_1 is only present in u_0, whereas c_2 is only in u_1. It is thus necessary to distinguish between the resources available in the different use-cases.

Definition 11 For every use-case $u \in U$, each link $l \in L$ has a *residual slot table* $t_i(u, l)$. Additionally, the residual resources consist of the *capacity* not yet reserved $cap_i(u, l) \in \mathbb{N}$ measured in a number of slots, and a set of *link users* denoted $use_i(u, l) \in \mathcal{P}(C)$.

The slot tables are used to store the reservations of already allocated channels, and thus hold the same information as alc_i, only organised differently. The latter serves to answer the question what slots are *allocated* to a specific channel, whereas the slot tables are used in evaluating potential paths and thus have to answer what slots are *available* to a channel. In addition to the residual slot table of a link, the capacity captures the throughput requirements of all link users. While this information might seem redundant in the presence of the residual slots, we do not only use it to reflect what is reserved by already allocated channels, but also channels that are not yet allocated, but where either the source or the destination port is mapped to an NI. When a port is mapped to an NI, there is only one ingress and egress link that has to accommodate the requirements of the channel. Consequently, we use the capacity and link users to reflect what channels are or will be using a link. We discuss this further when the allocation algorithm is introduced. Similar to the mapping and allocation function, the residual resources change over iterations and are therefore subscripted with an index. Initially, $\forall u \in U$ and $\forall l \in L$, $cap_0(u,l) = s_{\text{tbl}}$, $use_0(u,l) = \varnothing$ and $t_0(u,l) = \langle \varnothing, \ldots, \varnothing \rangle$. That is, the capacity of all links is s_{tbl}, no link has any users, and all slot tables are empty.

The functions introduced in Definition 11 enable the allocation algorithm to evaluate individual links. When allocating paths, however, looking at the slot table of a single link is not enough. Due to the requirements on slot alignment imposed by the pipelined virtual circuits, the slot tables along a path together determine what slots are available. To facilitate evaluation of an entire path we thus introduce the *path slot table*.

Definition 12 A path $\phi = \langle l_1, \ldots, l_k \rangle$ has a *path slot table* $t_i(u,\phi)$ where every residual link slot table $t_i(u,l_j)$, $j = 1..k$, is shifted cyclically $j-1$ steps left and a slot in $t_i(u,\phi)$ is the element-wise union of all shifted tables, for the specific iteration i and use-case u.

Consider, for example, the path $\phi = \langle l_1, l_2 \rangle$ traversed by c_1 in Fig. 4.8a where $t_i(u_0,l_1) = \langle \{c_0,c_3\}, \{c_0,c_3\}, \{c_1\} \rangle$ and $t_i(u_0,l_2) = \langle \{c_1\}, \varnothing, \varnothing \rangle$. After cyclically shifting the slot table of l_2 one step left, giving us $\langle \varnothing, \varnothing, \{c_1\} \rangle$, we arrive at $t_i(u,\phi)$ being $\langle \{c_0,c_3\}, \{c_0,c_3\}, \{c_1\} \rangle$.

Until this point, we have only looked at residual resources for individual use-cases. Channels, as we have seen, typically span multiple use-cases. Therefore, when allocating a channel c in an iteration i, the resources available on a specific link l are determined by looking not at one use-case, but rather at all use-cases in U_c.

Definition 13 The *aggregate slot table* $t_i(c,\phi)$ is an element-wise union across the use-cases in U_c. The *aggregate capacity* $cap_i(c,l)$ is defined as $\min_{u \in U_c} cap_i(u,l)$, i.e. the minimum capacity across the use-cases. Finally, the *aggregate link users*, denoted $use_i(c,l)$, is defined as $\bigcup_{u \in U_c} use_i(u,l)$, i.e. the union across the use-cases.

Now it is time to put Definitions 1 through 13 into practice and look at how they all come together in the resource allocation algorithm.

4.3 Allocation Algorithm

With the bus ports allocated, it remains to map the mappable ports to NIs, and to select paths and time slots so that the channel requirements are fulfilled. Two important requirements can be identified and the onus is, in both cases, on the unified allocation algorithm. First, the latency and throughput requirements of individual channels must be satisfied. Second, all channels must fit within the available resources.

Algorithm 4.3.1 Allocation of all channels C

1. **let** the set of unallocated channels $C'_0 := C$
2. **while** $C'_i \neq \varnothing$:
   a. **let** c denote the most critical channel (Section 4.3.1)
   b. **let** $C'_{i+1} := C'_i \setminus [c]$
   c. **for** every channel in $[c]$
      i. remove speculative capacity reservations (Section 4.3.2)
      ii. select a constrained path (Section 4.3.3)
      iii. refine the mapping function (Section 4.3.4)
      iv. find a slot allocation (Section 4.3.5)
      v. reserve resources (Section 4.3.6)
3. connect the control ports of the buses according to the mapping

As already mentioned, we use a heuristic algorithm where the mapping process is combined with the path selection and slot allocation, resulting in quick pruning of the solution search space [70]. The core of the proposed algorithm for mapping of multiple applications onto a network is outlined in Algorithm 4.3.1 and introduced here, after which further explanations follow in Sections 4.3.1 through 4.3.6.

The body of the allocation algorithm is iteration over the monotonically decreasing set of unallocated channels C'_i. In every iteration, we choose the most critical channel that has not yet been assigned a path and time slots, and potentially has source and destination ports not yet mapped to NIs. Once a channel is selected, a path (that also determines the NI mapping [70]) is selected, based on the resources that are available in all use-cases in which this channel is present [72]. After selecting a path, the mapping is refined if the source or destination is a mapping-group node. On this path, a set of time slots are selected such that throughput and latency requirements are met. Thereafter, the allocation of the channel is updated and the residual resources of the use-cases it spans are refined. To mitigate resource fragmentation, the allocation steps are repeated for all the channels in a tree before selecting a new channel [71]. If the channels in a tree are not allocated consecutively, chances are that the slots leading to minimal fragmentation are already reserved. The procedure is repeated until all channels are allocated. As further discussed in Section 4.3.7, we never backtrack to re-evaluate an already allocated channel. This results in low time complexity at the expense of optimality (or even failure to accommodate all channels). Once all channels are allocated, we connect the control ports of the initiator and target buses to the control buses of the NIs. We now discuss the different steps of the allocation algorithm.

4.3.1 Channel Traversal Order

The allocation algorithm is based on greedy iteration over the channels, and the order is therefore of utmost importance. The order should reflect how difficult it is to satisfy the requirements of a certain channel, and simultaneously aim to reduce resource fragmentation. Therefore, the order in which channels are allocated, i.e. their criticality, takes four different measures into account. In order of priority:

1. The number of use-cases that the channel spans $|U_c|$.
2. The number of channels in the channel tree $|[c]|$.
3. The latency requirement $\theta(c)$.
4. The throughput requirement $\rho(c)$.

Note that the selection of the most critical channel is done across all applications in parallel. Next, we discuss the different metrics and their ordering.

The first measure asserts that channels spanning many use-cases are allocated early. This ordering aims to reduce resource fragmentation as an application spanning several use-cases may only use time slots that are available in *all* the use-cases. Hence, for our example system, the channels of the control application are allocated first (with $|U_c| = 6$), followed by the status application (having $|U_c| = 4$), irrespective of the temporal requirements of any channel. The second measure is similar to the first in that channels in a tree may only use the resources available in *all* the branches. In conclusion, the first two measures sort the channels based on the constraints they impose on resource availability, rather than the requirements of the individual channels.

The third and fourth measures are different from the first two in that they are concerned with the temporal requirements of the channels. In contrast to many other network-allocation algorithms, we do not look only at throughput, but also at latency requirements. Moreover, as seen in Definition 1, the latency requirement is given as an absolute time rather than a number of hops. The criticality prioritises latency over throughput since the former places requirements both on the number of slots and their distribution, whereas the latter is only concerned with the number of slots. We return to discuss the details of the slot allocation in Section 4.3.5.

The channel traversal order highlights two important points. First, due to the sharing of time slots between mutually exclusive applications, the use-cases must be taken into account to avoid resource fragmentation. This is also the case for mutually exclusive channels within an application, i.e. in a channel tree. Second, latency requirements, that are often overlooked in network resource allocation, place more restrictions on the resource allocation than throughput requirements.

4.3.2 Speculative Reservation

Once a channel c has been selected, the next step is to prepare for the path selection by restoring potential capacity reservations. When the source port $src(c)$ of a channel c is mapped to an NI, the communication burden placed on the ingress and egress links of the NI is not necessarily determined by c alone. This is due to the mapping groups, as all ports in $[src(c)]$ are fixed to the same NI. Hence, all channels emanating from those ports must be accommodated on the egress link of that NI.

Similarly, the ingress link of the same NI has to accommodate all channels incident to those ports. Just as for the source port, when the destination port $dst(c)$ is mapped to an NI, all channels to or from the mapping group $[dst(c)]$ are forced to use that same NI.

The aforementioned issue arises due to the fact that channels are *interdependent*, but allocated independently, thus failing to recognise that the channels are linked via the mapping groups of their source and destination ports. Failing to acknowledge the above may result in overallocation of network resources. Numerous channels, still not allocated, may be forced to use the ingress and egress link due to a mapping that is already fixed. An NI may thereby be associated with a set of implicit requirements, not accounted for when evaluating possible paths. Ultimately this may cause the allocation to fail. Rather than attempting to allocate multiple channels simultaneously, we choose to incorporate the knowledge of the interdependent channels (sharing a port group) through speculative reservations.

Reservations are made after a channel is allocated, during the refinement of the mapping, as described Section 4.3.4. These reservations are then taken into account when evaluating paths in Section 4.3.3. If the mapping of either source or destination port is refined, we speculatively reserve capacity on the ingress and egress link of the corresponding NI for all interdependent unallocated channels. We do not know what paths these channels will eventually occupy, but the ingress and egress link is fixed at the moment the port is mapped to an NI. At this point we also have no knowledge of exactly what time slots are needed by the interdependent channels. Instead, we estimate the capacity by looking at the throughput requirements of the channels.[3]

Algorithm 4.3.2 shows how ingress capacity is reserved when a port p is mapped to an NI, i.e. $map_i([p]) \in N_n$. The first step is to identify the unallocated channels that have a source port in the same mapping group as p (potentially even the same port in the case of channel trees). Then, for every such channel and every use-case that channel is in, we see if it is already using the ingress link. Due to the channel sets it is possible that another channel already made a reservation for the entire set. If not, a reservation is made. Similarly, for egress reservations, Algorithm 4.3.3 looks

---

Algorithm 4.3.2 Ingress capacity reservation for port p

1. **for** all unallocated channels $c \in C'_i$ where $src(c)$ is in $[p]$

   a. **for** all $u \in U_c$, if $c \notin use_i(u, l)$, where l is the ingress link of $map_i([p])$

      i. **let** $cap_{i+1}(u, l) = cap_i(u, l) - \rho(c)$
      ii. **let** $use_{i+1}(u, l) = use_i(u, l) \cup [c]$
      iii. **if** $cap_{i+1}(u, l) < 0$ **fail**

---

[3] The latency requirement is hence only considered in the path evaluation and slot allocation. However, as a pre-processing step, we adjust the throughput requirement to also reflect the minimum number of slots required to satisfy the latency requirement.

Algorithm 4.3.3 Egress capacity reservation for port p

1. **for** all unallocated channels $c \in C_i'$ where $dst(c)$ is in $[p]$
   a. **for** all $u \in U_c$, if $c \notin use_i(u, l)$, where l is the egress link of $map_i([p])$
      i. **let** $cap_{i+1}(u, l) = cap_i(u, l) - \rho(c)$
      ii. **let** $use_{i+1}(u, l) = use_i(u, l) \cup [c]$
      iii. **if** $cap_{i+1}(u, l) < 0$ **fail**

at channels where the destination ports are in the same group as the newly mapped port.

For both ingress and egress reservations, it is possible that the capacity of a link goes below zero as the reservation is performed. In those cases, the algorithm (implemented in C++) throws an exception, thereby allowing the outer loop of the allocation algorithm to take appropriate measures. We return to discuss how such exceptions are handled, and the potential failure of the allocation algorithm in Section 4.3.7.

The speculative reservations enable us to incorporate knowledge of yet unallocated channels in the path selection. When a channel is about to be allocated, however, those speculative reservations must first be removed to reflect what resources are actually available prior to the allocation. Thus, if the source or destination port is already mapped to an NI, speculative reservations are removed from the ingress link in Algorithm 4.3.4 and egress link in Algorithm 4.3.5.

Comparing Algorithm 4.3.4 with Algorithm 4.3.2, it is clear that the former performs nothing but the inverse operations. Algorithm 4.3.5, in restoring egress reservations, looks at the destination rather than the source node of the channel, and the egress rather than ingress link of the corresponding NI. Other than these two points, the algorithms for restoring ingress and egress reservations are identical.

Algorithm 4.3.4 Remove ingress capacity reservation for c

1. **if** $src(c) \in P_i'$
   a. **for** all $u \in U_c$, if $c \in use_i(u, l)$, where l is the ingress link of $map_i([src(c)])$
      i. **let** $cap_{i+1}(u, l) = cap_i(u, l) + \rho(c)$.
      ii. **let** $use_{i+1}(u, l) = use_i(u, l) \setminus [c]$

Algorithm 4.3.5 Remove egress capacity reservation for c

1. **if** $dst(c) \in P_i'$
   a. **for** all $u \in U_c$, if $c \in use_i(u, l)$, where l is the egress link of $map_i([dst(c)])$
      i. **let** $cap_{i+1}(u, l) = cap_i(u, l) + \rho(c)$.
      ii. **let** $use_{i+1}(u, l) = use_i(u, l) \setminus [c]$

It is worth noting that in case a mapping group has a set of eligible NIs with only one NI, the mapping is in fact known, and speculative reservations are possible, even before the allocation starts. This is, for example, the case for the most of the port groups in our example system in Fig. 4.7. In our current implementation of the allocation algorithm we accommodate this knowledge by preceding the allocation of channels with a refinement of the mapping for all groups q where $|\epsilon(q)| = 1$.

4.3.3 Path Selection

The path selection is a central part of the allocation algorithm, as it determines the mapping, the path through the network and valid time slots. While fulfilling the requirements of the individual channel, the path should also use as few resources as possible to improve the likelihood of a successful allocation for all remaining channels. Three things set our problem apart from general constrained path selection [38]:

- All channels and requirements are known up-front.
- We have to consider the alignment of slots along the path.
- The residual resources depend on the use-case.

An in-depth discussion of these differences follows.

First, the knowledge of future requirements is an important parameter when selecting a path for a new channel as it enables better planning and management of resources [65]. In our path-selection algorithm, knowledge of future channels is incorporated through the speculative reservations that are used both as path constraints, and as part of the cost function. Future (unallocated) channels that have their source or destination port mapped are thus accounted for, albeit only by means of an estimation of their required capacity. We return to discuss the details of the cost function and path constraints in further depth when the algorithm is presented.

The second difference with existing path-selection algorithms, where constraints such as hop count and throughput are common [38, 195], is that slots on consecutive links must be aligned. Hence, it is not only a matter of slot availability on a link-to-link basis, as this fails to reflect the temporal dependencies introduced by the pipelined virtual circuits [70]. Consider, for example, a situation where a channel c_4 arrives at the router in Fig. 4.8a through the northern ingress link. If we look only at residual capacity for the individual links, c_4 prefers the link going east (two available slots) over the one heading south (one available slot). However, if c_4, due to an already traversed path ϕ, is limited to use the first slot on the egress link, e.g. with a path slot table $t(u_0, \phi) = \langle\{c_5\}, \{c_5\}, \varnothing\rangle$, then south is actually a better choice than east, since going east would leave us with a slot table with no available slots, a clearly unfeasible path.

Another problem related to the slot alignment is that alternative paths to the same intermediate node may result in widely varying sets of available slots. Thus, in contrast to traditional relaxation-based path-optimisation methods such as Dijkstra [48] or Bellman-Ford [16, 53], the best partial path is not necessarily the one with lowest cost [38]. Due to the alignment of slots on future links, a low cost path might eventually lead to a violation of the constraints, making a higher cost (but feasible) partial path beneficial.

The third difference is that the availability of resources depends on the channel being allocated, and what use-cases that channel spans. When evaluating paths it is thus necessary to consider what applications may run in parallel with a channel that is currently being allocated, and what resources are already reserved by the channels in those applications. We return to discuss the three differences and how they affect the path selection as the different parts of the algorithm are presented.

The outer loop of the path selection is seen in Algorithm 4.3.6. The search is based on A*Prune [110], continuously evaluating the intermediary least-cost path rather than the intermediary least-cost node [48]. Compared to the algorithm proposed in [70], this allows us to store a number of paths to each intermediate node, thus addressing the aforementioned issue of slot alignment. The search begins at the node where the source of the channel is currently mapped. The algorithm starts by determining which time slots are available even before the first link.[4] This is done by pruning all slots occupied by other channels in *any* of the already allocated branches. Consider for example the allocation of c_C in Fig. 4.6b, with c_A and c_B already allocated. Only the slots that are available on both paths are considered for c_C. In every iteration of Algorithm 4.3.6 we see if the last link in the path leads to the node where the destination of the channel is mapped. If so, and the path contains more than one link, then the selection is complete and the path returned. The reason we require the path length to be more than one link is to prevent paths going directly from a mapping node to an NI.[5] A path must never go via a mapping-group node (only start at the source and end at the destination) and must incorporate at least one physical network link (even with the source and destination mapped to the same NI). Note that the path check cannot be incorporated in the pruning process as a partial path must be allowed to go through the NI, and then leave it again before returning.

Algorithm 4.3.6 Selection of a path for channel c

1. determine the available slots t based on $[c]$
2. visit $map_i([src(c)])$ with the preceding path $\langle\rangle$ and slot table t
3. **while** there are paths in the path priority queue

    a. **let** ϕ denote the least cost path
    b. **let** n denote $dst(\text{last } \phi)$
    c. **if** $n = map_i([dst(c)])$ and $|\phi| > 1$ **return** ϕ
    d. **else** visit n with the preceding path ϕ

---

[4] For converging channel trees we swap the source and destination and perform the path selection backwards. This is to facilitate the derivation of available slots.

[5] The other way round is prevented by the link pruning in Algorithm 4.3.7.

Algorithm 4.3.7 Visit a node n with a preceding path ϕ

1. **for** every egress link l of n
   a. **if** $dst(l) \in N_q$ and $dst(l) \neq map_i([dst(c)])$ then reject l
   b. **else if** $l \in \phi$ then reject l
   c. **else if** $c \notin use_i(c, l)$ and $\rho(c) > cap_i(c, l)$ then reject l
   d. **else if** $t_i(c, \phi)$ cannot accommodate $\theta(c)$ and $\rho(c)$ then reject l
   e. **else** append l to ϕ and push on the path queue

The check thus rejects paths that are not valid as final paths (but are perfectly fine as partial paths). If the destination is not found or the path is not valid, the search continues from the current node onwards by visiting this node with the current path preceding it.

The top-level selection algorithm relies on pruning for an efficient traversal of the design space. Path pruning is done when visiting a node, and is crucial for the speed of the algorithm and quality of the solution.[6] In addition to the criteria for pruning, the cost function plays an important role in deciding which candidate path is preferred. Next we describe the pruning criteria, followed by the path cost function.

Algorithm 4.3.7 describes the procedure of visiting a node. As seen in the algorithm, we have four pruning criteria that a link has to pass, appearing in the given order due to the time it takes to evaluate them (fast to slow). First, to avoid traversing links that cannot possibly lead to valid path, we only visit a mapping-group node if it is the destination node. This rule improves the speed of the path selection by not exploring parts of the design space that contain unfeasible solutions. Second, to avoid loops, we do not traverse links that are already part of the path. There are situations when a path loop might be positive for the slot alignment (using the network links as intermediate storage), but we choose to limit the search space in favour of a faster path selection. Third, if the link is not already used by the channel (due to an interdependent channel), the aggregate residual capacity must be sufficient to accommodate the throughput requirement. Here we see that the knowledge of other channels is incorporated through the capacity reservations. Moreover, due to the formulation of the capacity function, the rule takes the multiple use-cases into account and guarantees that the link has the required residual capacity in all use-cases spanned by the channel in question. The last pruning criterion, also related to the channel requirements, is based on time-slot availability, similar to what is proposed in [70]. We determine the available slots on the partial path, and prune the link if the intermediate slot table cannot accommodate the throughput and latency requirements. The fourth rule addresses the problem of slot alignment by considering the link given the already traversed path. The aggregate slot table function also takes the multiple use-cases into account by merging the tables across the use-cases

[6] The pruning also makes it easy to incorporate path restrictions, e.g. to avoid deadlock. This is used when allocating resources for best-effort connections in Æthereal.

of the channel. Thus, if the path selection is able to find a set of slots that fulfil the latency and throughput requirements in the aggregate slot table, then these slots are available in all use-cases of the channel. Note that the check does not perform the actual slot allocation. Instead, it is a fast evaluation based solely on the residual slots and the maximum slot distance.

With the criteria for pruning in place, what remains for the path selection to be complete is the cost function that determines the order in which paths are evaluated. As shown in Definition 14, we include both slot availability and resource consumption in the cost measure by using a linear combination of hop count and channel contention, as suggested in [106]. Minimising hop count is beneficial as it conserves network resources, i.e. time slots on the links, and reduces power consumption [86, 181]. In addition to hop count, we incorporate contention by taking the maximum of two different measures. The first measure is the estimated average load (actual capacity minus reserved speculative capacity), similar to what is proposed in [94].[7] The differences in residual resources for the use-cases is taken into account by using the aggregate capacity function, returning the minimum capacity available in any of the use-cases of the channel being allocated. Since capacity reservations are only used on the ingress and egress links of the NIs, the second measure in the max expression is the reduction in the number of available slots incurred by the last link, i.e. the difference between what is available after $\langle l_1, \ldots, l_{k-1}\rangle$ and $\langle l_1, \ldots, l_k\rangle$, respectively. Thus, when allocating a channel tree, the link cost is based on how much capacity is available (including what is already reserved for the tree). This serves to maximise the overlap between the individual channels, for example by choosing the path $\star\star$ instead of $\star\star\star$ for channel c_C in Fig. 4.6b. The coefficients serve to control the normalisation and importance of the contention and distance measures, and they enable the user of the compile-time flow to influence the order in which paths are evaluated.

Definition 14 When allocating a channel c, the cost of a path $\phi = \langle l_1, \ldots, l_k\rangle$, with a partial path $\phi' = \langle l_1, \ldots, l_{k-1}\rangle$, is determined by $cost(\phi) = cost(\phi') + \alpha_{cap} \max(s_{\text{tbl}} - cap_i(c, l_k), |\sigma(t_i(c, \phi'))| - |\sigma(t_i(c, \phi))|) + \alpha_{dist}$ where $cost(\langle\rangle) = 0$ and the two coefficients α_{cap} and α_{dist} control the importance and normalisation of the cost contributions.

With the cost function in place, the path-selection algorithm is complete. Before moving to the mapping refinement, however, a few properties are worth highlighting. As seen in Algorithms 4.3.6 and 4.3.7, the inclusion of the mapping-group nodes has minimal impact on the path-selection algorithm, that unknowingly is taking mapping decisions. Similarly, through the formulation of aggregate capacity and aggregate slot table, the complexities of multiple use-cases and slot alignment are taken into account without changing the basic path-selection procedure.

---

[7] The authors suggest selecting of paths that interfere least with future channels through a heuristic called *Minimum Interference Routing Algorithm (MIRA)*. The algorithm does not only consider the ingress and egress links but also calculates an interference metric for every intermediate link in the network.

4.3.4 Refinement of Mapping

Once a path ϕ is selected for a channel c, Algorithm 4.3.8 checks whether $src(c)$ is not yet mapped to an NI. If not, the first link in ϕ (from a mapping-group node to an NI) decides the NI to which the port group is to be mapped and the current mapping function is refined with the newly determined mapping to a node in N_n. This refinement affects every port in $[src(c)]$ that is now in P'_{i+1}. When the mapping is updated, we also reserve ingress and egress capacity for the source on the NI in question. This is done according to Algorithms 4.3.4 and 4.3.5.

Similar to the source mapping refinement in Algorithm 4.3.8, we perform the corresponding update for the destination in Algorithm 4.3.9. If the destination of c is mapped to a mapping-group node, then we update the mapping function with the source node of the last link in the path (from an NI to a mapping-group node). As a last step, we also reserve ingress and egress capacity on the NI for the destination port of the channel.

After the refinement step, the ports in $[src(c)]$ and $[dst(c)]$ are all mapped to NIs, i.e. they are in P'_{i+1}. Moreover, all channels incident to or emanating from any of these ports are either allocated ($c \notin C'_i$) and have permanent capacity reservations along the entire path given by $alc_i(c)$, or are yet unallocated ($c \in C'_i$) in which case they have speculative capacity reservations on the ingress and egress links of the two NIs.

Algorithm 4.3.8 Refine mapping for source of c

---

1. **if** $src(c) \notin P'_i$

   a. **let** $map_{i+1} = map_i \oplus \{[src(c)] \mapsto n_n\}$ where n_n is $dst(\text{head } \phi)$
   b. reserve ingress and egress capacity for $src(c)$ on n_n

---

Algorithm 4.3.9 Refine mapping for destination of c

---

1. **if** $dst(c) \notin P'_i$

   a. **let** $map_{i+1} = map_i \oplus \{[dst(c)] \mapsto n_n\}$ where n_n is $src(\text{last } \phi)$
   b. reserve ingress and egress capacity for $dst(c)$ on n_n

---

4.3.5 Slot Allocation

After the path selection and refinement of the mapping function, we reach the point where all the efforts spent in performing reservations and evaluating candidate paths are to be put to the test when attempting to find a set of time slots such that the latency and throughput requirements of the channel are fulfilled. There are multiple challenges that the slot allocation algorithm has to address. First, the algorithm must deliver both latency and throughput guarantees. This translates to constraints both on

Algorithm 4.3.10 Allocate slots for a channel c on a path ϕ

1. **let** $t = t_i(c, \phi)$
2. **if** $\sigma(c, t) = \varnothing$ **fail**
3. **let** $\mathit{first} = \min \sigma(c, t)$
4. **let** $T_0 = \{\mathit{first}\}$
5. accommodate the latency requirement $\theta(c) - |\phi|$
6. accommodate the throughput requirement $\rho(c)$

the distances between slots and the number of slots reserved. Second, the throughput requirements must take the insertion of packet headers into account. That is, a slot does not always carry the same amount of payload data.

As seen in Algorithm 4.3.10, we start by determining what resources are already reserved by finding the aggregate slot table of the path. The allocation immediately fails if no slot is available. Next, we determine the first available slot in the aggregate table, *first*. In addition, the first slot is put in the set of taken slots, T, where we store the slot allocation of the current channel. After the initialisation, we proceed by accommodating the scheduling latency of the channel, that is, the latency requirement of the channel after subtracting the latency contribution of the entire path. The latter is, due to the pipelined virtual circuits and absence of contention, directly proportional to the length of the path. Allocation of the latency requirement is done according to Algorithm 4.3.11, followed by allocation of the throughput requirement, as described in Algorithm 4.3.12. We now discuss the two algorithms in detail.

Algorithm 4.3.11 Accommodate the latency θ for a channel c

1. **let** $\mathit{step} = \lfloor \theta \rfloor$
2. **if** $\mathit{first} > \mathit{step}$ **fail**
3. **let** $\mathit{previous}_0 = \mathit{first}$
4. **let** $\mathit{current}_0 = \mathit{first}$
5. **while** $\mathit{first} + s_{\mathrm{tbl}} - 1 - \mathit{previous}_j \geq \mathit{step}$
   a. **if** $\mathit{previous}_j + \mathit{step} > s_{\mathrm{tbl}}$
      i. **let** $\mathit{current}_k = s_{\mathrm{tbl}} - 1$
   b. **else**
      i. **let** $\mathit{current}_k = \mathit{previous}_j + \mathit{step}$
   c. **while** $\mathit{current}_k > \mathit{previous}_j$ and $\mathit{current}_k \notin \sigma(c, t)$
      i. **let** $\mathit{current}_{k+1} = \mathit{current}_k - 1$
   d. **if** $\mathit{current}_k = \mathit{previous}_j$ **fail**
   e. **let** $T_{j+1} = T_j \cup \{\mathit{current}_k\}$
   f. **let** $\mathit{previous}_{j+1} = \mathit{current}_k$

The scheduling latency of a channel is determined by the distance between reserved slots.[8] Thus, to guarantee an upper bound on the latency, is necessary to find a set of slots such that the distance between reserved slots is below a certain maximum value. A lower latency bound reduces the allowed distance, and consequently increases the number of slots required. This clearly illustrates the inverse coupling between throughput and latency incurred by the TDM arbitration.

The latency allocation algorithm, as shown in Algorithm 4.3.11, starts by determining the maximum allowed *step*, measured in a number of slots. The step conservatively rounds the latency requirement down. If the first slot is already larger than the step, then the allocation fails. Then, the algorithm iteratively evaluates if the gap between the first slot in the succeeding slot-table revolution and the previously allocated slot is bigger than or equal to the maximum step. If the two slots are further than *step* apart, then the allocation is not yet complete and we enter the loop that starts by determining where to continue the search. In the general case, we simply move one *step* forward by incrementing *current*. Should that slot be outside the first revolution of the table, then the search continues from the last slot of the slot table. The inner loop performs a linear search towards the previously allocated slot (by reducing *current* by one), thus ensuring that the maximum distance is strictly smaller than the step. If no available slot is found before encountering the previously allocated slot, the allocation fails. Once a slot is found, it is placed in the set of T slots, and the *previous* slot is updated. After Algorithm 4.3.11 completes, T holds the allocated slots and the overall allocation in Algorithm 4.3.10 moves on to accommodate the throughput requirement.

Figure 4.9 illustrates the slot allocation of a channel with a latency requirement of 4.3 slots. The first available slot is slot 3, and the step is conservatively rounded down to 4. After slot 3, the algorithm encounters slot 7 that is unavailable. Going towards the previously allocated slot, the linear search stops at slot 6, as it is the first available slot. From slot 6, we do not add the entire step. Instead, we go to the last slot of the table, and continue the search at slot 9. As this slot is available, it is added to the taken slots, thereby completing the latency allocation.

With the latency accommodated, it remains to ensure that enough slots are allocated to meet the throughput requirement. The throughput is determined by the slot table size, the number of slots allocated, the flit size, but also by the number of packet headers required. As described in Chapter 3, the header insertion is determined by the reservation of the preceding slot. That is, if the preceding slot is empty or allocated to another channel, then a new packet starts, and a header is inserted. In addition, to avoid starvation of the end-to-end flow control credits, the architecture enforces a maximum number of consecutive slots before headers are forcefully inserted.

Allocation of the throughput requirement, as described in Algorithm 4.3.12, continues where the latency allocation finished by determining the capacity allocated after accommodating the latency requirement. The requirement as well as the capac-

[8] We return to discuss the latency and throughput of a channel in Chapter 6.

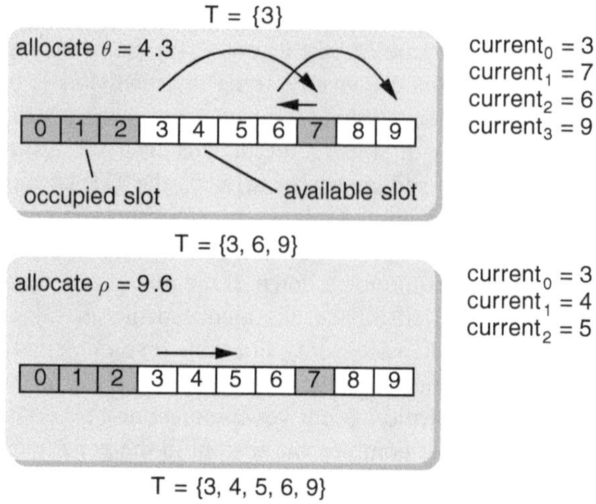

Fig. 4.9 Slot allocation

ity is measured in words per slot table revolution. For every slot that is already in the set of taken slots, we conservatively add the amount of payload data of a header flit. Next, the algorithm, which is based on linear search, starts from the first available slot. As we search through the set of available slots, we record how many consecutive slots we reserve by incrementing *counter*. The value of *counter* is then used to determine the payload size of the next slot. If the current slot is not available, then we reset the counter to reflect that a header must be inserted in a succeeding flit. In the case of an available slot, we see if it is already reserved during the latency allocation, i.e. if it is in T. If the slot is indeed taken, then we deduct the conservative capacity added during the initialisation of Algorithm 4.3.12. During

---

Algorithm 4.3.12 Accommodate the throughput ρ for a channel c

1. **let** $capacity_0 = |T|(s_{\text{flit}} - s_{\text{hdr}})$
2. **let** $current_0 = first$
3. **let** $counter_0 = 0$
4. **while** $capacity_j < \rho$ and $current_j > s_{\text{tbl}}$
   a. **if** $current_j \notin \sigma(c, t)$ **let** $counter_{j+1} = 0$
   b. **else**
      i. **if** $current_j \in T$ **let** $capacity_{j+1} = capacity_j - (s_{\text{flit}} - s_{\text{hdr}})$
      ii. **else let** $T_{j+1} = T_j \cup current_j$
      iii. **if** $counter_j = 0$ **let** $capacity_{j+1} = capacity_j + s_{\text{flit}} - s_{\text{hdr}}$
      iv. **else let** $capacity_{j+1} = capacity_j + s_{\text{flit}}$
      v. **if** $counter_j = s_{\text{pkt}} - 1$ **let** $counter_{j+1} = 0$
      vi. **else let** $counter_{j+1} = counter_j + 1$
5. **if** $capacity < \rho$ **fail**

---

the initialisation we are forced to assume that the slot carries a header, but it might now be in a block of consecutive slots. If the slot is not yet taken, then we add it to T. Next, the payload words are added to the capacity, based on whether the current slot is the first one in a group or not. Finally, the counter is updated based on a comparison with the maximum packet size. After the loop completes, we compare the allocated capacity with the requested throughput and the allocation fails if the requirement is not satisfied.

Continuing our example in Fig. 4.9, the throughput allocation, that has to accommodate 9.6 words of payload data, starts by deriving how much capacity is already reserved during the latency allocation. In this case T contains three slots, which gives us $capacity_0 = 6$ (assuming $s_{\text{flit}} = 3$ and $s_{\text{hdr}} = 1$). From slot 3, we then begin the linear search. For slot 3, that is already taken, we remove $s_{\text{flit}} - s_{\text{hdr}}$ from the capacity, only to add it again as $counter_0 = 0$. Moving to slot 4, the slot is added to the set of taken slots, and we add s_{flit} to the reserved capacity since $counter_1 = 1$ (assuming $s_{\text{pkt}} = 4$). After the addition of slot 4 $capacity_2 = 9$, which is still smaller than the requirement. The throughput allocation continues by evaluating slot 5, and adds this slot as well. An additional s_{flit} words are added to the reserved capacity and the allocation stops. The final set of slots, $T = \{3, 4, 5, 6, 9\}$, have a maximum distance of 3 slots and allow 13 words to be sent in one slot table revolution.

After executing Algorithms 4.3.11 and 4.3.12, Algorithm 4.3.10 completes and the slots in T are sufficient to accommodate the latency and throughput requirement of the channel or one of the algorithms failed during the allocation. In case of failure, the slot allocation throws an exception, indicating whether the latency or throughput allocation failed. If the slot allocation is successful, the set of slots in T are passed on to the resource reservation.

4.3.6 Resource Reservation

As the last step of the allocation algorithm, it remains to update the allocation and residual resources to reflect the decisions taken during the path selection and slot allocation. Algorithm 4.3.13 starts by refining the allocation function. Then, the reservation continues by traversing the path, and for each link the slot table, capacity and link users are updated for all the use-cases of the channel. Note that the reservation is done for all channels in $[c]$. The capacity reservation is similar to Algorithms 4.3.2 and 4.3.3. In contrast to the speculative reservations, the update of the link capacity is not using the throughput requirement of the channel (tree), but rather the number of reserved slots. Before removing the capacity, we remove any previous capacity reserved by the channel (tree). This capacity is determined by looking a the number of slots previously reserved to the channel. For every link along the path, the slot table of the link is merged with the newly reserved slots (without any conflicts between channels). After each link, the reserved slots are incremented by one, reflecting the pipelined nature of the channels.

Similar to the speculative reservation, Algorithm 4.3.13 might fail due to insufficient link capacity, $cap_{i+1}(u, l) < 0$, after performing the reservation. In such cases,

Algorithm 4.3.13 Reservation of path ϕ and slots T for a channel c

1. **let** $alc_{i+1} = alc_i \oplus \{c \mapsto (\phi, T)\}$
2. **for** $c' \in [c]$ and $u \in U'_c$ and $l \in \phi'$, determined by the path of $alc_{i+1}(c')$
   a. **if** $l =$ head ϕ' or $l =$ last ϕ'
      i. **if** $c' \in use_i(u, l)$
         A. **let** $cap_{i+1}(u, l) = cap_i(u, l) + |$set of slots of $alc_i(c')|$
      ii. **let** $cap_{i+1}(u, l) = cap_i(u, l) - |T|$
      iii. **let** $use_{i+1}(u, l) = use_i(u, l) \cup [c]$
   b. **let** $t_{i+1}(u, l) = t_i(u, l)$ with T reserved to $[c]$
   c. add one (modulo s_{tbl}) to every element in T

an exception is thrown indicating which link causes the problem. If the reservation is performed without violating any link capacity, the allocation of c is complete and Algorithm 4.3.1 continues by selecting a new channel for allocation. The resource reservation completes our exposition of the allocation algorithm, and we continue by looking at the limitations of the proposed solution.

4.3.7 Limitations

Although the proposed compile-time allocation flow has many merits, it is not without limitations. Starting at the problem formulation, our definition of an application allows independent starting and stopping of applications, but it does not allow for any variation within applications. That is, an application with multiple modes [60] (or scenarios) is either considered a single application or multiple independent applications. In the former case it might be necessary to over-dimension the requirements to those of the most demanding mode. In the latter case, on the other hand, the connections are stopped and started on mode changes, leading to disruptions in the provided service. It is possible, however, to provide bounds on the transition, as we shall see in Chapter 5. Moreover, the modes can be grouped to minimise the number of disruptions, similar to what is proposed in [138], but here on the application level.

The proposed algorithms are all heuristics that might fail even if a solution exists. If the algorithm is executed at design time and the application requirements cause the allocation to fail, it is possible to extended the network topology, similar to what is proposed in [12]. The potential extensions include more NIs and thus fewer channels per ingress and egress link, alternatively more routers or more links between routers and thus fewer channels per link inside the router network. It is also possible to loosen the mapping constraints, and thereby divide the requirements of different connections across more NIs. Similarly, the sets of eligible NIs can be extended to allow more flexibility in the mapping. The designer can also steer the algorithm by defining stricter sets of eligible NIs for the mapping groups. Increasing the slot table size helps reducing discretisation effects and thus helps in satisfying throughput requirements. As a last option, an increased interconnect clock frequency simplifies the task of finding an allocation. All of the aforementioned options are only possible at design time. At compile time, the allocation is *pass or fail*. That is, if a set of allocations are found, then they satisfy the requirements and there are no late surprises. Should the allocation fail, the only option is to change the use-case constraints or requirements of individual channels. Back tracking has been shown to improve the success rate of the allocation algorithm [188], but is currently not part of our implementation.

4.4 Experimental Results

To evaluate the scalability of the proposed algorithm, and its ability to find feasible solutions, we apply it to a range of random synthetic test cases. We vary the number of IPs (strictly the number of ports), and evaluate systems with 16, 32, 64 and 128 memory-mapped initiators and targets. To accommodate the IPs, we use a 2×2, 2×4, 4×4, and 8×4 mesh, respectively. For all network topologies we use two NIs per router and a slot table size of 32 slots and an operating frequency of 500 MHz. All ports of an IP are placed in a mapping group, and the set of eligible NIs is not restricted. That is, the ports of an IP (as a set) can be mapped to any NI.

In addition to varying the number of IPs, we also vary the number of applications mapped to them. Each application has a random number of connections, selected based on a normal distribution with an average of 10 connections and a standard deviation of 5. The source and destination IPs of the connections are chosen randomly, but not uniformly. To reflect the communication patterns of real SoCs, with bottleneck communication, characterising designs with shared off-chip memory, a fourth of the IPs have a probability of being chosen that is four times higher than the others. For each connection with the applications, throughput and latency requirements are varied across three bins respectively (30, 300 and 3,000 ns, in combination with 3, 30, and 300 Mbps). This reflects for example a video SoC where video flows have high throughput requirements, audio has low throughput needs, and the control flows have low throughput needs but are latency critical.

As the last parameter, we vary the number of edges each application adds to the use-case constrains. With each edge, the size of the cliques grow and more applications are thus running in parallel. This translates to fewer use-cases, but also more contention and less opportunity to re-use resources between mutually exclusive applications.

We evaluate 100 benchmarks for each design point, and the value presented is the average of those runs.[9] In addition to the execution time, we also report the

[9] In fact, the execution time varies significantly. In some cases the standard deviation in execution time is in the order of half the reported mean time.

failure rate, that is, how many of the 100 runs failed. A failure occurs when no feasible solution is found. It is possible that no feasible solution exists for the given requirements or that a solution was not found. There is no attempt made to recover from a failed allocation, e.g. by relaxing constraints, adding links or increasing the slot table size. Similarly, we do not attempt to reduce cost in the case of a successful allocation by, e.g., using a smaller interconnect or reducing the operating frequency. Some of these techniques are, however, part of the larger design flow as shown in Fig. 1.8.

We start by looking at scalability for different topology sizes. The results in Fig. 4.10 show how the execution time and failure rate scales when varying the number of IPs and network sizes. Independent of the topology size, we consider four applications, and two constraint edges for each application. For the smaller system sizes, there is thus more contention, since the same IPs are more likely to be involved in multiple applications. This results in a failure to allocate 30% of the designs. For the larger system sizes, all allocations are successful. However, as clearly seen in Fig. 4.10, the execution time grows exponentially due to the path selection algorithm. For contemporary systems, with a few tens to hundreds of IPs, the time is still reasonable.

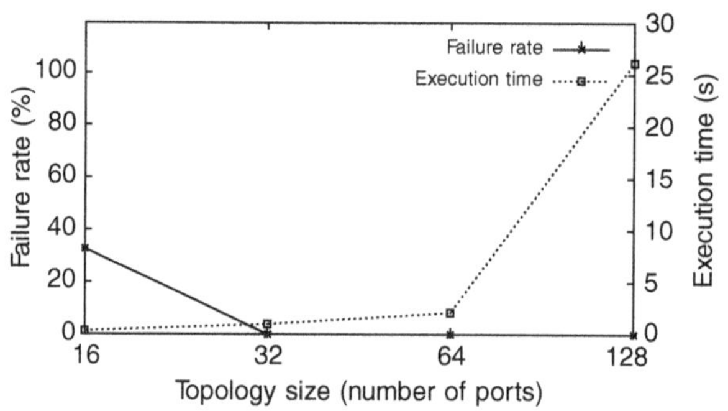

Fig. 4.10 Scaling with the number of IP ports and network size

Figure 4.11 shows the behaviour when scaling the number of applications, but leaving the number of IPs and the network topology constant. The system used for the experiments has 128 IPs, and independent of the number of applications one constraint edge is added for each of them. As seen in Fig. 4.11, every single allocation is successful for 2, 4 and 8 applications. When the number of applications reaches 16, however, almost 30% of the allocations fail. This is result of the increased contention for the shared bottleneck IPs, and the choice to force all ports of a specific IP to one NI (i.e. the grouping of ports). Figure 4.11 also shows how the execution time grows with the number of applications (and thus connections and channels). The foundation of the allocation algorithm is iteration over the channels, but for each channel we iterate over all unallocated channels when managing reservations. This suggest a quadratic growth in the number of channels, but the experiments suggest that the execution time grows roughly linearly with the number of applications.

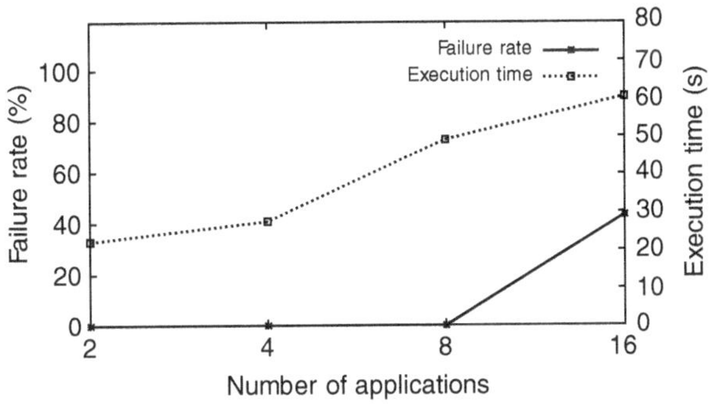

Fig. 4.11 Scaling with the number applications

Lastly, we fix both the number of IPs and applications, and instead vary the number of constraint edges. Figure 4.12 illustrates the failure rate and execution time for a system with 128 IPs and 16 applications. The first point, with one constraint edge per application, corresponds to the point with 16 applications in Fig. 4.11. As the number of edges in the constraint graph increase, the use-cases grow in size, and reduce in number. Figure 4.12 shows that the execution time and failure rate follow each other closely, and both increase with the amount of contention in the network. With fewer constraint edges, the allocation algorithm is able to exploit the mutually exclusivity and share resources between applications that are not part of the same use-case.

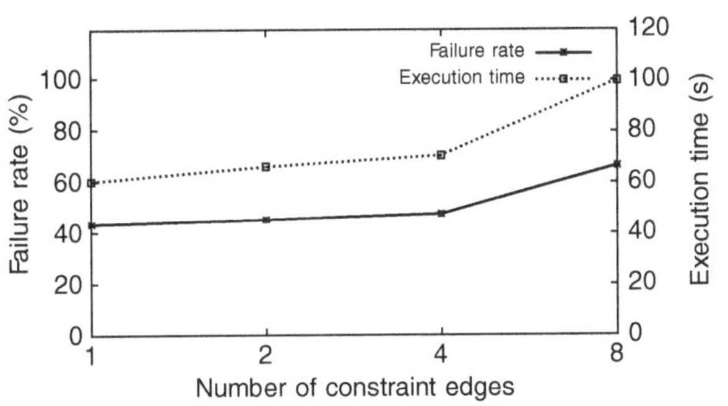

Fig. 4.12 Scaling with the number constraint edges

4.5 Conclusions

In this chapter we have presented the allocation part of the design flow, emphasising the problem formulation and the algorithms used to allocate resources. We summarise by evaluating the contribution to the different high-level requirements in Chapter 1.

Rather than contributing positively to *scalability*, the compile-time flow limits the scalability due to the complexity of the algorithms. Despite the use of heuristics, the core of the path selection is A*Prune which is known to have exponential growth in run time for certain problem instances [110]. The problem is mitigated by using a path-selection algorithm based on Dijkstra, as proposed in [70], but this on the other hand compromises the quality of the solution and might cause the allocation to fail. For contemporary systems, with hundreds of IPs and connections, the proposed algorithms are sufficient, but other approaches might be necessary as the problem size grows.

The resource allocation contributes to application *diversity* by not making any assumptions about how the resources are used. In contrast to e.g. [156, 188], the requirements of individual channels do not infer anything about the application behaviour, and can be worst-case or average-case requirements. Another important property for the diversity of the compile-time flow is that requirements are expressed as latency and throughput bounds, similar to well-established frameworks [185]. Requirements are thus specified on a level that can be understood by the application designer, who needs no knowledge about the interconnect implementation.

The compile-time flow contributes to *composability* through the formulations of use-cases as combinations of applications, and resource allocations on the level of channels. Given these, the algorithm allocates resources such that two concurrent applications never contend. Moreover, the compile-time flow has an important contribution to *predictability* as the channels are assigned resources that guarantees the fulfilment of their throughput and latency requirements. Additionally, the problem formulation and allocation algorithms ensures *reconfigurability* at the level of applications, by assigning resources such that applications can be started and stopped independently.

The most important contribution of this chapter is to the *automation* of the proposed interconnect. After dimensioning the interconnect, all the application requirements have to be translated into resource allocations. This translates into deciding what bus ports, NIs, paths and time slots to use, and the design space is truly enormous. Our proposed heuristic algorithm quickly prunes the vast number of infeasible options, by incorporating time-slot alignment in the path evaluation, and tightly coupling path selection and mapping of IPs to NIs. The compile-time flow corresponds to two different tools in the design flow, one for assigning bus ports and one for the mapping, routing and slot allocation.

Chapter 5
Instantiation

After completing the dimensioning and allocation, the interconnect can be instantiated. An instantiation consists of two parts. First, the hardware, in the form of SystemC models or HDL. Second, the software, comprising the resource allocations and the appropriate run-time libraries. With both the hardware and software in place, the entire interconnect can be simulated or synthesised. This chapter, corresponding to Step 3 in Fig. 1.8, takes its starting point in a complete specification of the architecture and resource allocation, and addresses the challenges involved in turning the specification into a hardware and software realisation.

To instantiate the hardware, the appropriate blocks must be created and interconnected. As we already know from Chapter 3, many blocks are parametrisable and require the appropriate values to be set. In addition to the instantiation of individual library modules, such as buses and routers, these modules must also be connected according to the architecture description. On the network and interconnect level we thus have to generate instance-specific hardware. The last part of the hardware instantiation is a set of auxiliary files that are required by the ASIC/FPGA tools for, e.g., HDL elaboration, simulation and RTL synthesis.

The hardware is of no use without the appropriate (host) software. To instantiate the latter, the first step is to translate allocations into source code. At run-time, the host then instantiates these allocations as a result of trigger events [145]. To do so, however, the host must know how to access the register files in the programmable hardware components, i.e. the NIs and buses. This requires detailed knowledge about the register layout of the individual components. Moreover, the configuration registers are distributed across the interconnect and the host must also be able to reach the individual registers.

In addition to the challenges involved in instantiating one specific allocation, the host must also be able to close connections. When modifying or closing connections, both the interconnect and the IPs must be left in a consistent state, that is, a state from which the system can continue processing normally rather than progressing towards an error state [102]. Simply updating the NI registers could cause out-of-order delivery inside the network, or even loss of data, with an erroneous behaviour, e.g. deadlock, as the outcome [68]

Lastly, besides correctness, certain applications require a bound on the time needed to start up, e.g. to guarantee a certain initialisation time, or a bound on

A. Hansson, K. Goossens, *On-Chip Interconnect with Aelite*, Embedded Systems, DOI 10.1007/978-1-4419-6865-4_5, © Springer Science+Business Media, LLC 2011

the transition time when switching between two closely related applications, e.g. two different decoders (applications) in a software-defined radio. This time includes opening (and closing) their connections. Beyond prior work, we enable upper bounds on execution time for the reconfiguration operations (with unavoidable assumptions on the IP behaviour).

As mentioned, the instantiation is divided into three parts, as illustrated in Fig. 5.1. We start by describing the hardware instantiation for simulation as well as synthesis (Section 5.1). We follow this by a discussion on how the allocations are translated into software (Section 5.2), and how together they are used by the host with the run-time libraries (Section 5.3). We end this chapter by presenting experimental results (Section 5.4) and conclusions (Section 5.5).

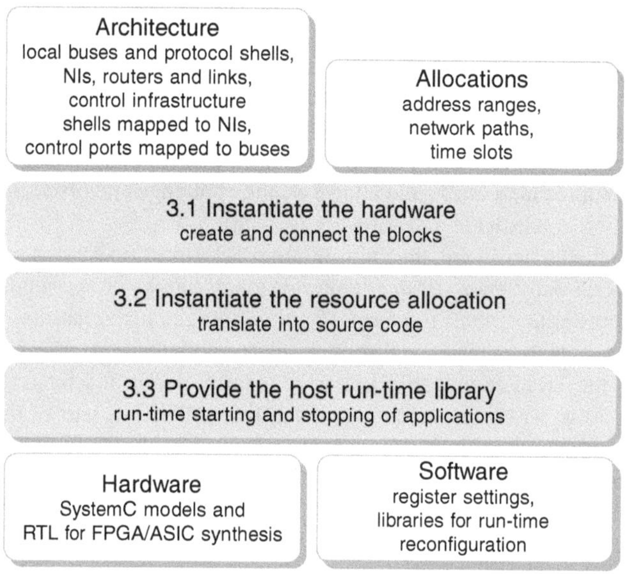

Fig. 5.1 Instantiation flow

5.1 Hardware

The end goal of the run-time instantiation is a RTL hardware description that can be turned into a physical implementation. Such a description is, however, inadequate for design space exploration due to low simulation speed. Therefore, the HDL description is generated together with a transaction-level SystemC model for each component. SystemC enables component descriptions with object-oriented design patterns, by offering a collection of interfaces and a simulation kernel. With a combination of both SystemC models and HDL descriptions, it is possible to refine descriptions from the functional to transaction, cycle-, bit-accurate and finally to gate level. Thus, very large scale integration and software implementations can proceed in parallel, as advocated in [166]. We first describe the instantiation of the SystemC model, followed by the hardware description of the interconnect.

5.1.1 SystemC Model

For SystemC simulation, we use run-time parsing and instantiation of the hardware (rather than compiling a fixed interconnect instance). Thus, when executing the simulator, as described in Chapter 6, the modules are created dynamically by calling the appropriate constructors. The dynamic creation allows easier exploration of different interconnect architectures (by simply changing the XML specification), at the price of slightly lower simulation performance. Depending on the type of module, different constructors are called, and the individual modules are responsible for their own parameter interpretation and port instantiation. This makes it easy to extend the SystemC model with new modules and relieves the simulator from knowing what information different modules need, and how they use it. Each instantiated module also registers its ports to make them known to the top-level that has to connect them. Once all modules are instantiated, the simulator connects the module ports according to the architecture specification.

To facilitate comparison between the different abstraction levels, there is a one-to-one correspondence between architectural units in the HDL and modules in the SystemC simulator. Furthermore, all naming of modules and ports is determined purely based on identifier fields in the XML and is thus consistent throughout SystemC and HDL. However, in contrast to the hardware description, the SystemC model does not make use of module hierarchies. Moreover, the two hardware instances differ greatly in speed and accuracy. In the SystemC model, the network is flit accurate on the inside and cycle accurate on the edges (NIs). The other components of the interconnect (buses, shells and atomiser) are cycle accurate. The interfaces to the IPs are either on the transaction level or cycle level, leaving this choice to the user of the SystemC model. Thanks to the higher abstraction levels, the SystemC model is roughly three orders of magnitude faster than bit-level HDL simulation and five orders of magnitude faster than gate-level netlist simulation. Additionally it offers more observability, with monitors for e.g. the temporal behaviour of connections and buffer levels in the NIs. We return to discuss these differences when evaluating application performance in Chapter 6.

Like the actual hardware description, the SystemC model of the interconnect requires run-time configuration. The SystemC model offers three different host implementations, depending on the purpose of the simulation. First, an *ubiquitous host* that does not use the control infrastructure, but instead makes use of the global observability and controllability (through pointers) to directly access the individual registers. The ubiquitous host carries out configuration operations instantaneously, and is useful when the user wishes to abstract from the time involved in opening and closing connections. Second, an *ideal host* where all computation is done without progressing the simulation time, i.e. infinitely fast. The configuration operations, however, are carried out using memory-mapped reads and writes over the actual control infrastructure. That is, computation is abstracted from, but the communication is modelled cycle accurate. Third, the *ARM host* implementation shown in Fig. 5.2, comprising a SWARM instruction-set simulator of an ARM7 [42]. The processor has a von Neumann architecture, and a target bus (with a static address decoder) within the tile determines whether the read and write transactions go to the local memory or to the interconnect. The external initiator port in Fig. 5.2 corresponds to the port on the host in Fig. 3.16a. The run-time libraries and the allocations are compiled and linked using a standard ARM tool chain (e.g. gcc or armcc) and automatically loaded to the local memory when instantiating the host tile.[1] In contrast to the ideal host, the ARM host executes the actual run-time library functions and thus includes both cycle-accurate computation and communication. In this model every instruction takes two cycles to fetch due to the bus address decoder and SRAM memory controller.

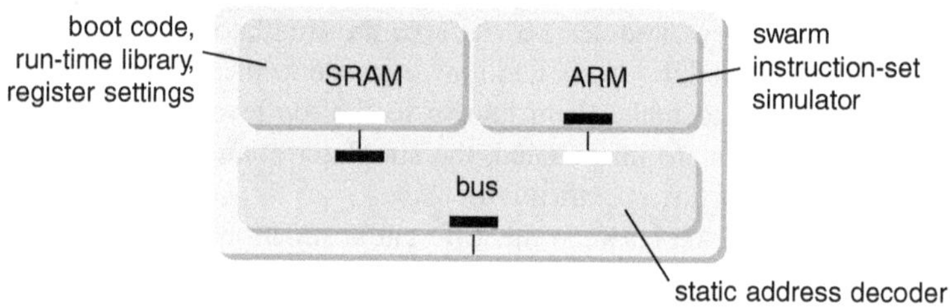

Fig. 5.2 ARM host instantiation

In addition to the interconnect and the host, the SystemC model also contains the IP modules, e.g. accelerators, other instruction-set simulators, and input/output peripherals. In the case a model of the IP is not available, library implementations of initiators and targets are provided in the form of traffic generators and memory controllers. We return to discuss the details of those in Chapter 6.

5.1.2 RTL Implementation

In contrast to the SystemC model, the instantiation of the interconnect RTL implementation is done by printing the HDL of the network and interconnect level, together with the appropriate support scripts for RTL synthesis. Once instantiated, the synthesisable HDL description of the interconnect can be used together with HDL simulators, as well as standard ASIC and FPGA back-end design flows.

The instantiation of the RTL implementation makes use of a limited number of parametrisable modules, as already introduced in Chapter 3. Bottom up, these

[1] In a real implementation this step requires more elaborate boot strapping, as discussed in Chapter 3.

modules are, link pipeline stages, routers, NIs, clock domain crossings, initiator and target shells, atomisers, and (programmable) initiator and target buses. Each of these modules has generics to control everything from the width of specific signal groups on a shell, to the register layout of the NI or the number of ports and the address map of a (non-programmable) target bus. For all the parametrisable modules, it is the responsibility of the run-time instantiation flow to assign the appropriate generics, based on the architecture specification.

The individual modules are instantiated in a design hierarchy with two levels. The first level contains only the network (NIs, routers and links) and is a structural description of the network topology. The interface of the network level is the ports of the NIs, i.e. the streaming ports and memory-mapped control ports. No flit ports are thus visible at this level. The second level surrounds the network with the remainder of the interconnect. Thus, both the network level and interconnect level are instance-specific structural descriptions of the architecture.

As already discussed, the hardware is not only design-time parametrisable, but also run-time configurable. Therefore, to use the interconnect RTL implementation, we also need to instantiate a host, similar to what is done in the SystemC model. In the RTL implementation we do not have the option of using a ubiquitous host. Instead we rely on a programmable exerciser, similar to the ideal host. This exerciser configures the interconnect using memory-mapped communication, without having to perform any computation. We also offer the option of using an ARM9 instruction-set simulator as the host, corresponding closely to the ARM host in the SystemC model. In the latter case the allocations and run-time library are loaded into the tightly coupled memories of the processor.

Like the SystemC model, the RTL implementation also includes behavioural models of traffic generators and memory controllers. The latter are used for all IPs for which no model or implementation is available. With the IPs, the hardware instantiation is complete, and we continue by introducing the first part of the software instantiation, namely the resource allocations.

5.2 Allocations

For the interconnect hardware to be useful, the resource allocations must be instantiated. While the task of orchestrating the instantiation falls on the run-time library, the functions of the library require information about the actual allocations, and also about the architecture of the interconnect. This information must therefore be translated into software, as part of the instantiation.

The translation of allocations to source code uses C structures for connections, and they in turn use structures for bus address-decoder settings and channels. Each channel also uses structures for the path and slot reservation. In the end, the allocations are translated into a collection of connection definitions, each one composed of nothing but unsigned integers and characters. The allocation of the virtual control infrastructure is instantiated like any other application.

Besides the resource allocations, the run-time library requires information about the instance-specific architectural constants. This includes the register layout of the NI and the flit header format. Additionally, the instantiation conveys the number of NIs and to which NI the host is connected. As we shall see next, the latter information is necessary to determine if an NI is local or remote when executing the operations of the run-time library.

5.3 Run-Time Library

Similar to [145, 168, 170, 176], the interconnect configuration is done by a general-purpose CPU and is initiated by events in the system, caused by, e.g., a user command to start or stop an application, an embedded resource manager [31, 104] or mode switches in the incoming streams [170]. The configuration of the interconnect is hidden from the application programmer by an application-level run-time API [144] to start and stop applications. Thereby, it is the responsibility of the configuration management system, not the user, to determine the specific ordering of the operations to apply [102] and guarantee the correctness and temporal behaviour of those operations. When starting and stopping applications, the opening and closing of individual connections is left to the interconnect run-time library software. The run-time library hides the underlying implementation and eases integration as well as modifications [111]. This allows for easy porting to a wide range of host implementations, such as the PC-based host used for the FPGA instantiation in Chapter 7. The final outcome is a set of calls to memory-mapped read and write operations, which is the only processor-specific part of the library.

The host controls the interconnect through three top-level operations:

1. initialise the control infrastructure,
2. open a new connection, and
3. close an existing connection.

The host functionality is implemented in portable C libraries [68, 77] to be executed on e.g. an embedded ARM or μBlaze processor, and three functions *art_cfg_init*, *art_open_conn* and *art_close_conn*, corresponding to the three top-level operations. The initialisation is performed as part of the bootstrap procedure, before any calls to the other functions. It instantiates the connections of the control infrastructure, from every remote NI back to the local NI such that the control infrastructure can be used for the interconnect reconfiguration.[2] Once the initialisation is complete, the last two operations open and close user connections, respectively. The opening or closing of a connection requires us to configure one NI on each side of the network, and potentially also an initiator bus and a target bus. The registers of those modules are not necessarily directly in the proximity of the host, so for each

[2] The initialisation is an optimisation. The most basic approach is to open and close the configuration connections every time they are used.

of these four locations, the general procedure is to open a control request channel from the host to the module in question, perform the programming, and then close the control request channel.

We now describe the three top-level operations, followed by a discussion on how to derive temporal bounds on the execution of the operations.

5.3.1 Initialisation

The control connections are opened in the initialisation phase, as illustrated in Fig. 5.3. The figure contains a subset of the architecture of the example system, and shows only the modules that are involved in the configuration of the interconnect, i.e. the buses, the NIs and the control infrastructure. The port names for the ARM and SRAM correspond to the mappable ports in Fig. 4.7, and now we also include the ports that are used for the control infrastructure. The latter are indexed with numerals rather than literals to distinguish between control ports and user ports.

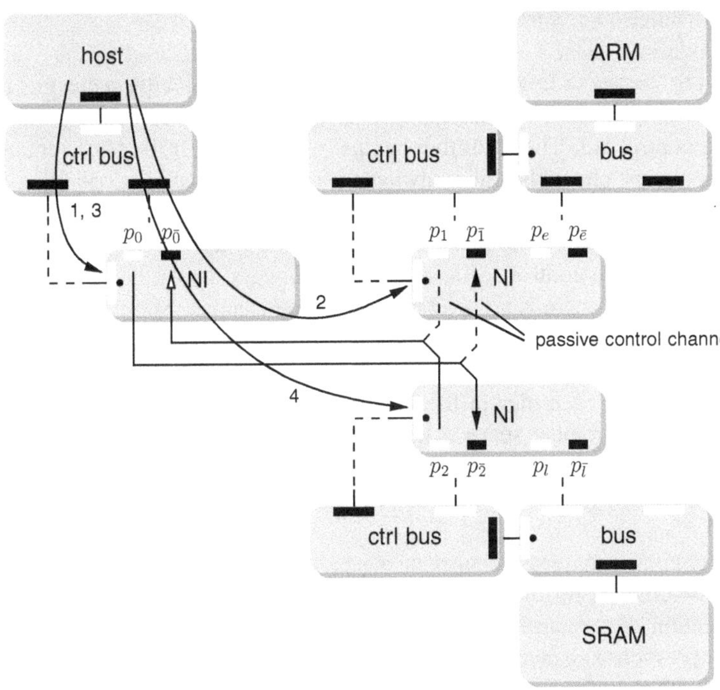

Fig. 5.3 Control connection initialisation

A control connection, like any other connection, consists of two channels. The control request channel connects the streaming target port on the host NI (p_0 in Fig. 5.3) to the initiator port on the remote NI ($p_{\bar{1}}$ and $p_{\bar{2}}$). Similarly, control response channels connects the target ports on the remote NIs (p_1 and p_2) to the

Algorithm 5.3.1 Open control request channel

1. **if** the current NI is remote
   a. **if** the current NI is not the active NI (initialised to the local one)
      i. **if** the active NI is remote, close the control request channel
   b. configure the path and the destination port identifier in the local NI
   c. **let** the active NI refer to the current NI

initiator port on the local NI ($p_{\bar{0}}$). The control request channels together constitute a channel tree and thus share their time slots. Moreover, all the buffers used for the control connections are uniformly sized across the remote NIs. It is thus possible to configure the remote space (initial amount of credits) and the slot reservation once, independent of the number of control request channels. Once the shared slots are reserved and the space is configured, we proceed by opening a control request and response channels to and from every remote NI.

The opening of a control request channel is done by the function *art_open_cfg_req* according to Algorithm 5.3.1. The control request channel makes the register in a remote NI accessible via the local control bus, the network, and the remote control bus. The algorithm begins by evaluating if the NI is remote. For the control request channels, this is always the case (as we call the operation for all the remote NIs), but for a general open or close operation the target NI might also be the local one. If the NI is indeed remote, we see if it is already connected. As seen in Fig. 5.3, one control request channel is always active, i.e. left unused but open. In these cases there is no need to re-open the channel. Should the current NI be different from the active one, the control request channel to the latter is closed. The final step in opening the control request channel is to configure the path and the destination port identifier in the local NI. As previously mentioned, the slot reservation and the remote space is initialised once and does not need any further configuration. Note that the opening of the control request channel involves only manipulation of registers in the local NI, as indicated by the arrows marked 1 and 3 in Fig. 5.3.

With the control request channel in place, we proceed by also opening a control response channel for the remote NI in question. The response channel enables communication of responses, e.g. the result of a read operation, back to the host. The control response channel is opened like any user channel is, according to Algorithm 5.3.2. That is, by writing to the registers of the remote NI, indicated by the arrows marked 2 and 4 in Fig. 5.3, the host sets the path, the destination port identifier, the remote space and the time slots. As a last step, the response channel is enabled for scheduling. For the response channels we also use a channel tree and thus share the time slots. However, even with the same slots being used on the links going to the host, each remote NIs has its own slot reservation since the number of hops (and thus the offset of the slots relative to the shared links) differ.

Due to the use of end-to-end flow control, the output buffer in the remote NI must be sufficiently large to fit three complete write commands (the minimum needed to execute Algorithm 5.3.2). If the output queue in the remote NI is smaller than this, the local NI runs out of credits before the remote NI is able so send any credits back, and the configuration (and entire system) deadlocks. In contrast to the user-specified connections, it is possible to let the connections of the control infrastructure not use end-to-end flow control. Not using end-to-end flow control makes it possible to close a control request channel as soon as the data is sent. Not having any flow control also removes the aforementioned constraint on the buffer sizes. However, not using flow control pushes the responsibility of timely consumption onto the control bus of the remote NI and the register files it is connected to. If these modules do not consume data equally fast as it arrives, e.g. due to multiple clock domains, configuration data is lost, with an undefined behaviour as the result. This is clearly unacceptable, and we therefore choose to use end-to-end flow control, and pay the price in time and buffering.

Algorithm 5.3.2 Open a channel

1. set the path and the destination port
2. set the initial space to the size of the output queue
3. set the time slots
4. set the enable flag

The response channels are left unused, but open, when moving to the initialisation of other remote NIs. This is illustrated in Fig. 5.3 by the passive control channels going to the NI of the ARM. Hence, in contrast to the request channels, the control response channels are set up *once* and are in place and ready to be used when opening and closing user-specified connections. As we have already seen in Algorithm 5.3.1, the control request channels are closed for every remote NI by calling the function *art_close_cfg_req*. Closing a channel, as later discussed in Section 5.3.3, involves waiting for all data to be sent, and any potential credits to return. These two conditions ensure that the connection has reached a *quiescent state* and that it is safe to reconfigure the interconnect. We return to discuss quiescence on the interconnect and IP level in Section 5.3.3. Before detailing how to close channels, however, we look at how channels are opened.

5.3.2 Opening a Connection

Once the initialisation is complete, we proceed to open the user-defined connections. Figure 5.4 shows the opening of a connection from the ARM to the SRAM. The first step is to establish a control request channel to the remote NI of the target bus (arrow 1). Once this channel is in place, the address decoder of the bus is configured, using the function *art_set_addr_dec* (arrow 2). Next, we proceed to open the first

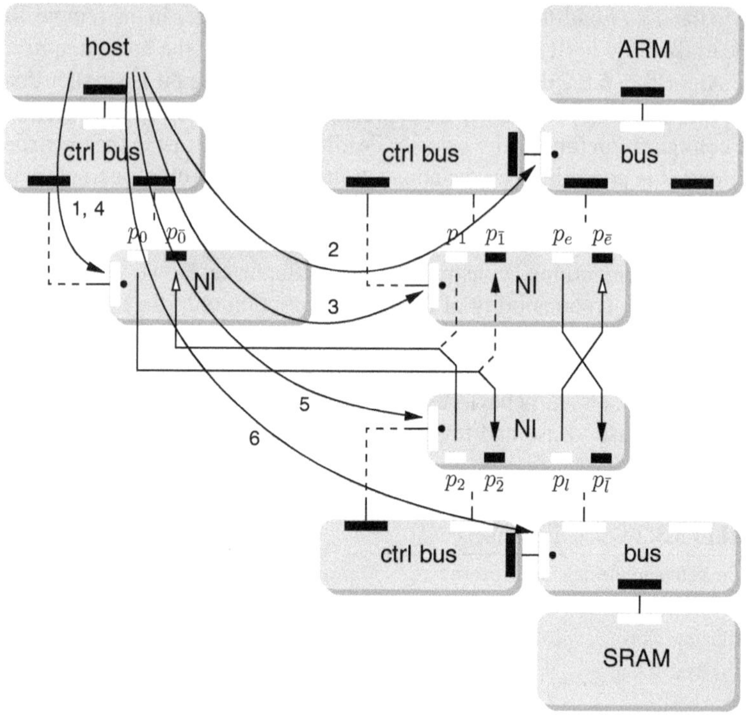

Fig. 5.4 Opening a user-specified connection

user channel, in this case the request channel going from the ARM to the SRAM (p_e to $p_{\bar{l}}$). The specific NI is already active, and we do not have to open a control request channel before opening the request channel according to Algorithm 5.3.2 (arrow 3). We have already seen this algorithm being used during the initialisation, and now we apply it to the user channel.

Algorithm 5.3.2 in turn calls a number of internal functions for encoding of the structure fields, e.g. *art_create_path*, and for writing to the specific registers, e.g. *art_set_flowctrl* and *art_set_slots*. Once the request channel from the ARM to the SRAM is opened, the configuration request channel to the NI of the ARM is closed and the NI of the SRAM becomes the active one (arrow 4). The opening of the connection continues by opening the response channel, also according to Algorithm 5.3.2 (arrow 5). The final step involves the configuration of the initiator bus in front of the SRAM (arrow 6). Also here we make use of the fact that the control port of the bus is connected to the active NI. There is thus no need to close and open any configuration request channel before configuring the bus. After the configuration of the initiator bus, the connection between the ARM and SRAM is opened and can be used. The connection remains in this state until it is closed by the host.

5.3.3 Closing a Connection

Algorithm 5.3.3 shows the procedure for closing a channel. When comparing with the algorithm for opening of a channel (Algorithm 5.3.2), we see that the first step is to await quiescence before conducting any actual reconfiguration. An opened channel is possibly in use, and the step serves to ensure that no information is lost. We return to discuss quiescence after presenting the remaining steps of Algorithm 5.3.3. After quiescence is reached, the closing of a channel is essentially the opening in reverse. First we disable the port, and then the slot reservation is removed. Note that there is no need to reset the space counter. The value causes no harm, and is set again as part of the channel opening procedure. Similarly, it is not necessary to reset the path and destination port. A new channel using the same port simply overwrites the old values.

Algorithm 5.3.3 Close a channel

1. await quiescence
2. unset the enable flag
3. unset the slots

Closing a connection is more challenging than opening one, as we have to ensure that no information is lost, and that the IPs and the building blocks of the interconnect are left in a consistent state. Using the definition of [102], a module is quiescent if (1) it will not initiate new transactions, (2) it is not currently engaged in a transaction that it initiated, (3) it is not currently engaged in servicing a transaction, and (4) no transactions have been or will be initiated which require service from this module. The same requirements apply not only to transactions in memory-mapped communication, but also to elements of streaming data and packets. When quiescence is reached on both the IP and interconnect level (all three stacks), a change can take place with a consistent state as the outcome.

The interconnect has no knowledge about potential outstanding transactions on the IP level. Furthermore, it is possible that an IP is involved in transactions with multiple other IPs, and that completion of one transaction depends on other transactions with other IPs [102]. Hence, quiescence on the IP level must be implemented by the IPs themselves. Starting at the ARM in Fig. 5.4, the processor must not initiate any new transactions, but still accept and service transactions that are outstanding. In terms of memory-mapped transactions, this translates to:

1. No new requests must be initiated, but ongoing request (e.g. a write burst) must finish.
2. All initiated transactions that require a response must be allowed to finish.

For an ARMv5 instruction-set processor the steps can be implemented using a memory barrier operation. Similar functionality is offered by the VLIW used in our example instance in Chapter 7. However, it is possible that such operations are not available, e.g. in the SWARM that only implements the ARMv4 instruction-set architecture. The memory mapped initiator (in this case the ARM host) is then under the impression that (posted) requests finish at the moment the interconnect accepted them. The aforementioned Step 1 thus requires the interconnect to also deliver the request to the final destination. Moreover, the availability of barrier operations relies on the memory-mapped protocol stack. Thus, for a streaming initiator we only assume that it offers functionality to stop injecting data and the responsibility of delivery is pushed to the interconnect.

When the initiator is quiescent, it still remains to ensure quiescence in the interconnect. Before quiescence is reached, all streaming data must be delivered to the destination. Hence, it must leave the input queue traverse the network, and be consumed from the output queue. In addition to delivering any outstanding data, also credits must be delivered (in both directions). That is, they must leave the source NI, traverse the network, and be added to the space counter in the destination NI. Algorithm 5.3.4 ensures that the aforementioned conditions are fulfilled.

Algorithm 5.3.4 Await quiescence

1. wait until the input queue is empty
2. wait until no credits are present
3. wait until the remote space reaches its original value

In best-effort networks that do not employ end-to-end flow control, quiescence in the router network, has to be implemented by, e.g., inserting a special tagged message as an end-of-stream marker [145], thus requiring the co-operation of the IPs. Thanks to the guaranteed services, and the presence of end-to-end flow control, we know that all streaming data relating to the channel in question has left the network when Algorithm 5.3.4 completes. As the input queue is emptied, the space counter is decremented accordingly. After this step, the data is either in the router network, in the output buffer, or already consumed by the target. It is not until all data is consumed that all credits are sent back. Thus, the last condition conservatively ensures that all streaming data belonging to the channel has left the network. Algorithm 5.3.4 on its own does not ensure that credits that are sent have reached the destination NI (only that they leave the source NI). The algorithm is therefore applied also to the channel going in the reverse direction.

The steps in Algorithm 5.3.4 are carried out by repeatedly polling memory-mapped registers in the source NI. One single word (32 bits) returns one bit per port indicating if the input queue is empty, if the credit counter is zero, and if the remote space is equal to what it was last programmed to (as part of the channel opening). For this purpose, the register file for space is mirrored, and only one copy is decremented and incremented when flits are sent and credits returned, respectively.

5.3.4 Temporal Bounds

The time required to perform the different control operations, also taking the opening and closing of control connections into account, is determined by three terms. First, the time the host requires to issue the programming operations, capturing the time needed to calculate or read the allocation from background memory. As we shall see in Section 5.4, the host has a major impact on the time required for reconfiguration. Second, the time required program the control registers, determined by the resources allocated to the virtual control infrastructure, and the amount of data communicated. Third and last, the time before the IP is quiescent, which greatly depends on the transactional state of the IP.

Starting with the first and second term, both the host implementation, the run-time library functions, and their use of the virtual control infrastructure can be captured in one of the models proposed in Chapter 6. Given the path and time slots assigned to a control connection, it is thus possible to derive an upper bound on the time required to open the control request channel, configure the remote NI or bus, and close the control request channel. The challenge is to capture the host implementation, i.e. the execution of the run-time library instructions on a processor. Once this is accomplished, the host is connected to the network via a protocol shell, just like any other IP, and the posted write transactions that implement the control operations are thus modelled similar to what is proposed in Chapter 6.

The third term in the reconfiguration time depends on the IP. In the general case, the time until a connection is quiescent cannot be bounded for a close operation. Closing the channel without awaiting a quiescent state is possible, but requires some guard functionality in the IP. This requires knowledge about the low-level details of the interconnect and complicates the IP design and separation of the protocol stacks. Therefore, we do not consider it in this work. Instead, we assume that it is possible to bound the time required to reach a quiescent state by inserting *reconfiguration points* [145] in the applications. With such functionality offer by the IPs, the total reconfiguration time can be conservatively bounded.

5.4 Experimental Results

To evaluate the scalability and predictability of the run-time library, we instantiate a number of randomly generated applications, and study the temporal behaviour and memory requirements. Throughout our experiments, we assume a word width of 32 bits used by both the IPs and the interconnect. The network, IPs and ARM host (including the local bus and memory) are all operating at 100 MHz.

To assess the impact of the host execution time, we compile two different binaries for the ARM. First, one with the connection structures and the run-time library as described in this chapter. Second, one where the allocations are pre-compiled into plain read and write calls, thus removing all computation and minimising the amount of function calls. Both binaries are compiled using *arm-elf-gcc* version 3.4.3 with the compiler options *−mcpu=arm7−Os*.

5.4.1 Setup Time

The first measure we study is the time required to initialise the control connections and how this grows with the number of (remote) NIs. Figure 5.5 shows the effect of increasing the mesh size from 1×4 to 5×4, constantly having two NIs per router. It is clear that the control infrastructure is not the limiting factor, as the ARM is consistently more than 2 times slower than the ideal host for the read/write implementation. The slow down stems from the 2 cycle delay in instruction fetches. The library slows the ARM host down another 7 times, making it roughly 15 times slower than the ideal host. The slow down roughly corresponds to number of arithmetic operations needed for every load or store. All three implementations show a linear growth, as expected with a constant work per NI.

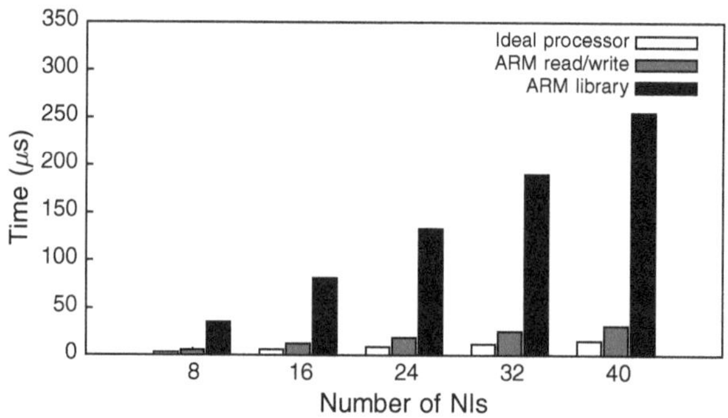

Fig. 5.5 Initialisation time for different number of NIs

Figure 5.6 shows the effect on the cumulative setup time when varying the number of connections on a fixed 4×4 mesh with two NIs per router. The connections are opened one by one according to Algorithm 5.3.2, with the control request channel opened and closed for every single channel. The setup time is measured from the point when initialisation of all the control response channels is completed. The

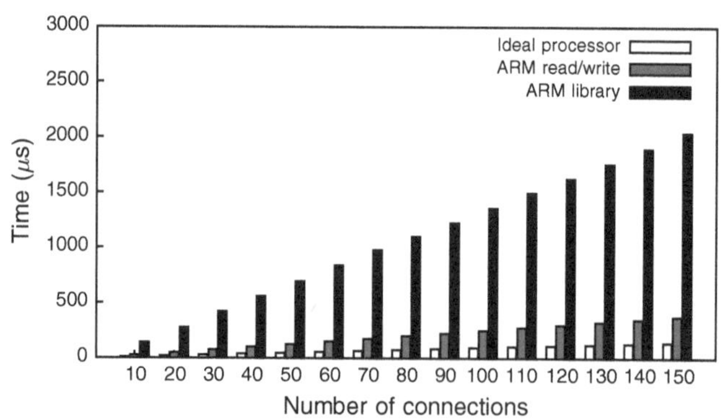

Fig. 5.6 Setup time for different number of connections

impact of the latter is shown in Fig. 5.5 (the scenario with 32 NIs). Also in this experiment the ARM is roughly 2 times slower than the ideal host when using read/write calls, and 15 times slower with the library implementation. Again, we see that the time required grows linearly with the number of connections.

We also evaluate the possibility to exploit locality when multiple channels share the same source NI and there are no ordering constraints between the channels, e.g. when opening connections. We iterate over the NIs rather than over the connections when opening a set of channels. This makes optimal use of Algorithm 5.3.1, as the active NI remains the same for as long as possible. Consequently, for every NI, the control request channel only has to be opened and closed once, reducing both the amount of programming required, and the time spent awaiting quiescence. After applying this strategy, the cumulative time still grows linearly with the number of connections, but at a much slower rate. For the ideal host, the time required to open all connections is less than half or what is shown in Fig. 5.6. Similarly for the ARM implementations, the total setup time is roughly 40% less using this technique.

The observed maximum-case setup times for a single connection are 1.09, 3.04 and 16.72 μs for the three implementations. Comparing with [68], where best-effort connections are used for the virtual control infrastructure, it is clear that the time introduced due to the infrastructure is much larger when using guaranteed services. The average time required to open a connection, for example, is reported to be 246 ns in [68], compared to almost 1 μs in this work. The fundamental difference, for which we pay a price in higher average case, is that worst-case bounds can be given. Using the analytical formulation, as proposed in Section 5.3.4, 1.41 μs is an upper bound on the reconfiguration time for the worst-case connection, which is allocated 2 out of 16 slots in the request direction and 1 slot in the response direction. The discrepancy with the measured 1.09 is partly due to the run-time conditions being less adverse than the worst case, but also to our choice of model (Fig. 6.5a), where pipelining is not taken into account.

5.4.2 Memory Requirements

Next, we assess the memory requirements of the ARM binary. To reduce the memory footprint, we also specify the bit widths of the different fields in the connection structure. Note though that bit members generally worsen the execution time as many compilers generate inefficient code for reading and writing them.

The binary size for a varying number of NIs is shown in Fig. 5.7. The various instantiations correspond to the same mesh networks as in Fig. 5.5, here together with a 40 connection use-case. Expanding the library functions to read and write calls roughly doubles the size. This is to be compared with the 15-fold speedup observed in Fig. 5.5.

Figure 5.8 shows the corresponding scaling with the number of connections. Here, the difference between the library and the read/write implementation becomes more obvious as the number of connections grows. As an example, using read and write calls instead of the library functions increases the size with 140%, from 18 kb

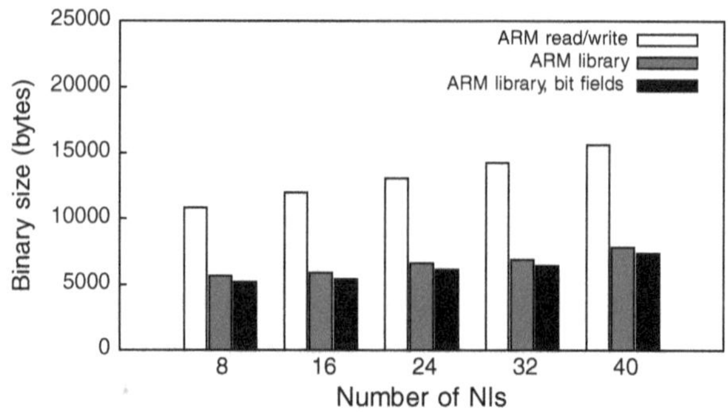

Fig. 5.7 Binary size for different number of NIs

to 43 kB, for the 150 connection case. From Fig. 5.8, we conclude that the effect of bit members is only a few percent reduction of the binary size. Moreover, as expected, we observe an execution time that is slightly worsened, although the measured increase is a mere 1.2%. Taking both these facts into account it is hardly justifiable to employ bit members unless memory footprint is absolutely critical.

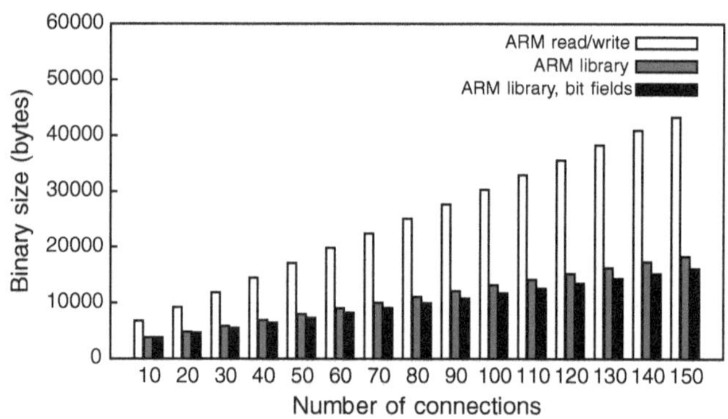

Fig. 5.8 Binary size for different number of connections

5.4.3 *Tear-Down Time*

Similar to Fig. 5.6, Fig. 5.9 shows the cumulative execution time required to tear down a use-case with a varying number of connections. Quiescence is enforced by instructing the IPs (traffic generators) to stop, and then using Algorithm 5.3.4. Hence, in contrast to the open operations where the time until a quiescent state is reached is zero, the tear down also involves waiting for in-flight transactions to finish. Here, we assume ideal IPs, where reconfiguration points occur between every single transaction. Even with this ideal model, the IP modules play an important part in determining the total time required to carry out a tear down operation. For

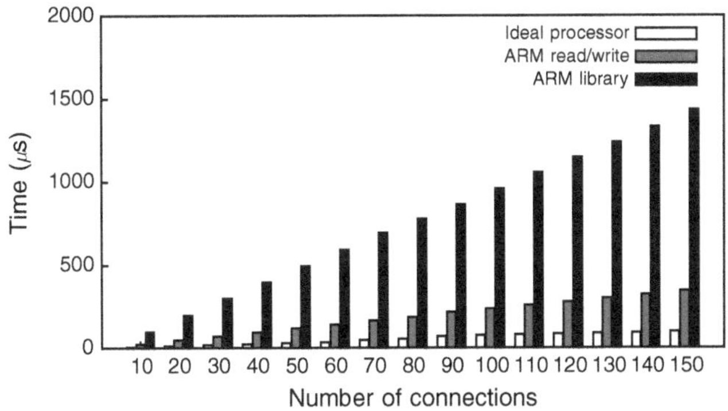

Fig. 5.9 Tear-down time

example, the measured worst case tear-down time is 12 μs for all three methods, due to a slow IP. The influence of the IP also causes a large variation across the different connections. For the ARM the fastest tear-down operation is completed in a little more than 1 μs while the library implementation requires 4.8 μs. The ideal host does the corresponding action in 12 ns, as the connection emanates from the local NI and only involves local reads and writes. Note that closing a channel (Algorithm 5.3.3) requires less programming than opening a channel (Algorithm 5.3.2). Thus, if the IPs is quiescent, closing is in fact faster than opening.

Table 5.1 summarises the analytical approximations of the linear terms in Figs. 5.5, 5.6, 5.7 and 5.8. The ARM library is roughly a factor six slower than the read/write calls, but has only half the binary size. Moreover, the library is reusable and can be compiled independent of the allocations. A combination of the two approaches is also possible. If the system frequently changes between a pre-defined set of applications, it is possible to pre-compute the instruction sequences for starting and stopping the applications in question, and thereby speedup the transition.

Table 5.1 Linear approximations of execution time and binary size

	Ideal host	ARM read/write	ARM library
Initialisation time / NI (ns)	375	784	6,385
Setup time / conn. (ns)	936	2,506	13,594
Tear-down time / conn. (ns)	852	2,305	9,575
Binary size / NI (bytes)	–	149	67
Binary size / conn. (bytes)	–	262	104

5.5 Conclusions

In this chapter we have presented the instantiation flow of the interconnect hardware and software. We summarise by evaluating the contribution to the different high-level requirements in Chapter 1.

The run-time instantiation, in its current form, limits the interconnect *scalability* as it uses a centralised host. In our evaluation, we see that the time required to reconfigure the interconnect is already substantial, and that it grows linearly with the size of the network and the amount of connections. However, the proposed run-time libraries do not inherently limit the scalability, and it is possible to distribute the work across multiple hosts. Different configuration operations can thus overlap in time, and the memory requirements are distributed across the hosts. In addition to the centralised host, the scalability is currently also limited in the SystemC model, as it unconditionally instantiates the entire system. The simulation speed thus degrades with the system size. To improve the scalability, it is possible to only instantiate the applications (IPs and interconnect components) of the current running use-case.

The contributions to *diversity* are many. First, the instantiation of the interconnect hardware and software uses industry-standard languages, and can thus be used with a wide range of target platforms and tools. Second, the hardware is instantiated both as transaction-level SystemC models and HDL RTL, to enable diversity in the speed an accuracy offered. Third, the dynamic hardware instantiation of the SystemC simulator offers easy addition of new modules. Lastly, the run-time library presents the user with a high-level view, facilitating diverse types of host implementations. Furthermore, the host libraries are provided as portable ANSI C code.

The run-time instantiation contributes to *composability* by performing the reconfiguration operations on the level of individual connections. Connections are thus started and stopped independently. The operations are also *predictable*, as the connections of the virtual control infrastructure are used to carry reconfiguration data. The challenge in providing temporal bounds on the reconfiguration operations lies in bounding the time required for the host to issue the operations, and in the case of closing, the time until the IPs are quiescent.

The main contribution of this chapter is the *reconfigurability* of the interconnect. Through a basic set of operations, the run-time library allows the user to dynamically open and close connections. The low level details of the interconnect are hidden, easing integration as well as modifications. Reconfiguration of the tasks running on IP modules [93, 144, 170] is an important task that lies outside the scope of this work.

Another important contribution of this chapter is the *automation* of hardware and software instantiation. Turning high-level specifications into HDL and C code is a tedious and error prone job. The run-time design flow greatly simplifies the task of the user by fully automating this process. We see examples of the automation in Chapter 7 where an entire system instance is created in a matter of hours.

Chapter 6
Verification

In the previous chapters, we have seen how the communication requirements of the applications are turned into a complete interconnect instance. The allocation of communication requirements, however, only covers the network and excludes the effects of end-to-end flow control. When NI ports are added as part of the dimensioning in Chapters 3 and 4, all buffers are given a default size, and until now we have assumed that the buffers are sufficiently large. If this is not the case, the application performance suffers, and conversely, if the buffers are unnecessarily large they waste power and area [28, 56, 79, 168]. In addition to the network, also the clock domain crossings, shells, atomisers and buses affect the performance and must be considered. Furthermore, the communication requirements are merely a result of higher-level application requirements, and it remains to verify that the applications have the expected functional and temporal behaviour when they are executed on the system instance, i.e. taking the interdependencies between the tasks, the IPs and the interconnect into account. Thus, for an existing interconnect instance, we must be able to *determine the temporal behaviour* of the applications mapped to it, given fixed NI buffer sizes. On the other hand, if we are designing an interconnect specifically for a set of applications, then it is desirable to *determine sufficiently large* NI buffers. In this chapter, which corresponds to Step 4 in Fig. 1.8, we show how to verify the application-level performance requirements (considering the entire interconnect) and how to size the NI buffers.

In the dimensioning, allocation and instantiation, the design flow does not distinguish between different types of applications. That is, all requirements are treated the same, whether they stem from a firm, soft or non-real-time application. The difference lies in what the requirements represent, e.g. worst-case or average-case throughput, and how they are derived by the application designer. In contrast to earlier steps in the design flow, here we need different approaches to verify the functional and temporal behaviour for different types of applications, and size their NI buffers accordingly. We first look at the wide range of options for verification based on *simulation*, and continue with the main contribution of this chapter: *conservative models of the network channels* that enable firm real-time guarantees.

Simulation, potentially based on statistical models [84], is a common approach to evaluate application performance [40, 135], and is the first step in our verification flow, depicted in Fig. 6.1. Using either the SystemC or RTL instantiation of the

A. Hansson, K. Goossens, *On-Chip Interconnect with Aelite*, Embedded Systems, DOI 10.1007/978-1-4419-6865-4_6, © Springer Science+Business Media, LLC 2011

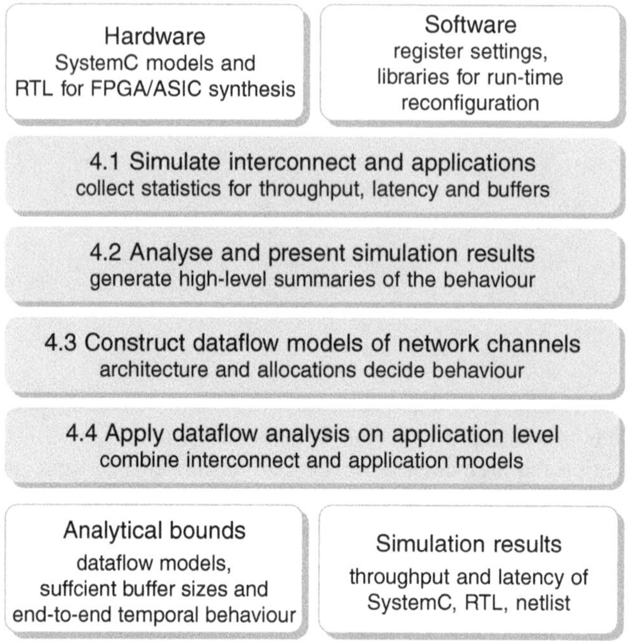

Fig. 6.1 Performance verification flow

interconnect, the designer can include the IPs and applications in the form of, e.g., transaction-level models, instruction-set simulators, or actual RTL implementations. In addition to any application-specific instrumentation, the interconnect provides extensive statistics concerning the throughput and latency of individual connections (for both SystemC and RTL simulation), together with the buffer usage in the NIs. Thus, it is possible to evaluate the performance and choose appropriate buffer sizes. As already discussed in Chapter 5, the instantiation flow also includes traffic generators and memory models, for those cases where no models of the IPs are available. The verification flow also presents high-level summaries of the simulation results. As part of the evaluation, it is also possible to compare the throughput and latency for the different levels of simulation, i.e. SystemC, RTL and netlist.

Simulation places *no restrictions on the applications* and is suitable for verification of applications with soft or no real-time requirements. However, simulation (in the general case) is not able to guarantee that the application requirements are met, even for the specific input trace. The composability of our interconnect ensures that the *observed behaviour is independent of other applications*. It is thus not necessary to consider variations due to changes in the input traces or behaviours of other applications in the system. This is a major qualitative difference with other (non-composable) interconnects and a crucial step in reducing verification complexity. Even with interference from other applications removed, however, a different starting time or uncertainty in the platform itself (e.g. due to a clock domain crossing) might cause variations in the application behaviour. As a result, the application might not deliver the expected performance, possibly even deadlock or exhibit timing related bugs when it is executed again, on the same system instance, with the same input trace. The inability to guarantee a certain temporal behaviour makes simulation an inappropriate technique for performance verification and buffer sizing for firm real-time applications.

The techniques proposed in [39, 56] overcome the limitations of simulation by modelling the application behaviour by means of a *traffic characterisation*. Thus, the work takes a network-centric approach and includes the IPs, buses, atomisers and shells in the characterisation. The work presented in [56] uses linear bounds to characterise traffic, while [39] assumes strictly periodic producers and consumers. NI buffer sizes are computed such that for every traffic source that adheres to the characterisation, there is always sufficient space in the buffer to allow data production. However, finding a traffic characterisation is difficult if not impossible as it is limited to point-to-point network channels. Thus, it is not possible to capture dependencies between different network channels, e.g. the synchronisation between channels in the target buses and initiator buses, or the dependencies between requests and responses for a memory-mapped target such as the SRAM. All these restrictions on the traffic characterisation severely limit the applicability of the methods. Moreover, neither of the approaches allow the availability of buffer space to influence the production time of data. Thereby, they are only able to compute buffer sizes given a temporal behaviour, and not the reverse. Hence, it is not possible to derive the temporal behaviour given fixed buffer sizes, i.e. to map a new application to an already existing interconnect instance.

As the main contribution of this chapter, we show how to construct a dataflow graph that conservatively models a channel of any network that offers guaranteed latency and throughput. As illustrated in Fig. 6.1, the generation of dataflow models is the third step in the verification flow. We exemplify the technique by constructing several models of the proposed network architecture. The applicability of the model is illustrated by using it together with state-of-the-art dataflow analysis techniques [15] to derive conservative bounds on buffer sizes in the NIs and temporal behaviour of the applications. This is the fourth and last step in the verification flow. If the verification flow fails, the communication requirements (the input to the design flow in Fig. 1.8) are adjusted accordingly, i.e. by increasing the requested throughput or decreasing the requested latency. In the case of a successful verification, the architecture description is (back) annotated with the minimal sufficient buffer sizes, and the interconnect is instantiated anew.

To evaluate our proposed channel model, we restrict our application model and compare computed buffer sizes with existing approaches [39, 56], for a range of SoC designs. Coupled with fast approximation techniques, buffer sizes are determined with a run time comparable to existing analytical methods [56], and results comparable to exhaustive simulation [39]. For larger SoC designs, where the simulation-based approach is infeasible, our approach finishes in seconds. Moreover, in Chapter 7 we demonstrate how the dataflow models enable us to capture the behaviour of both the application and the entire interconnect (not only the network) in one model, thus greatly improving the applicability compared to [39, 56].

The remainder of this chapter is structured as follows. We start by presenting the terminology and concepts of dataflow graphs together with their applications (Section 6.1). Next, we formulate the requirements on the interconnect (Section 6.2) and give a detailed description of the temporal behaviour of our proposed network architecture (Section 6.3), after which our proposed model of a communication channel is derived (Section 6.4). We compare the run time and buffer sizes derived using our approach with those of [39, 56] (Section 6.5). We end this chapter with conclusions (Section 6.6).

6.1 Problem Formulation

In this chapter we address the problem of verifying the performance requirements of the applications, and sizing of the NI buffers of the proposed interconnect. As discussed in Chapter 2, an end-to-end performance analysis requires us to conservatively characterise the behaviour of the applications, the behaviour of the IPs that run the applications, and the behaviour of the *entire interconnect*. We assume that the application models are given as variable-rate dataflow graphs [199]. Moreover, we assume that all hardware and software tasks, in the application as well as the architecture, only execute when they have input data available and sufficient space in all output buffers, i.e. they do not block during execution. In accordance with the proposed interconnect we assume blocking flow control.

In constructing a model of the interconnect we must consider that the behaviour of the target and initiator buses depend heavily on the IPs to which they are connected. For example, in Chapter 7 we show how to construct a conservative model for the target bus and SRAM in our example system, including detailed knowledge of the specific memory controller. Due to the difficulties in constructing a general model for the aforementioned parts of the interconnect, we restrict ourselves to constructing a model of the network. Thus, given a network architecture that offers guaranteed latency and throughput for individual channels, we must construct a channel model, as illustrated in Fig. 6.2, that conservatively models the network, and allows us to verify the application requirements and size the NI buffers. As seen in the figure, two instances of the model (one for each channel of a connection) conservatively captures the behaviour of a connection, from NI to NI. In Fig. 6.2, and throughout this chapter, dashed arrows are used to denote flow control (in this case the accept signal of the streaming interfaces as discussed in Chapter 3).

Once we have a dataflow graph of the interconnect and the application, it can be used to compute buffer capacities, to guarantee deadlock freedom, to guarantee satisfaction of latency and throughput constraints, and to guarantee periodic behaviour of sources and sinks [10, 15]. Figure 6.3 shows an example producer–consumer application, with the corresponding dataflow model on top. A producer task, e.g. a software task running on a processor, communicates via a buffer with a consumer task, e.g. the scheduler in an NI. If the buffer is full, the producer stalls, and the consumer stalls if the buffer is empty. The dataflow graph is divided in *components*

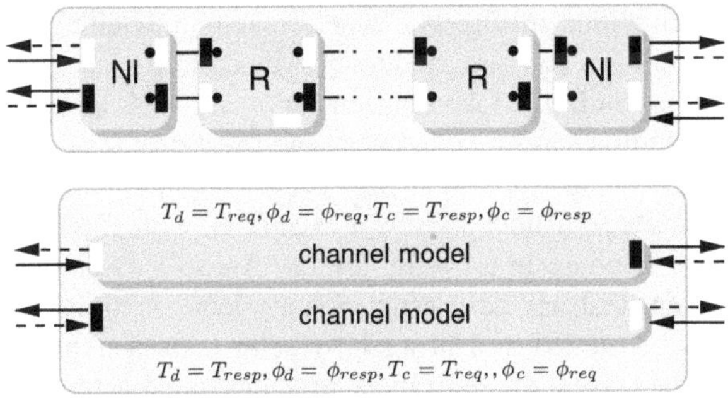

Fig. 6.2 Architecture and corresponding model

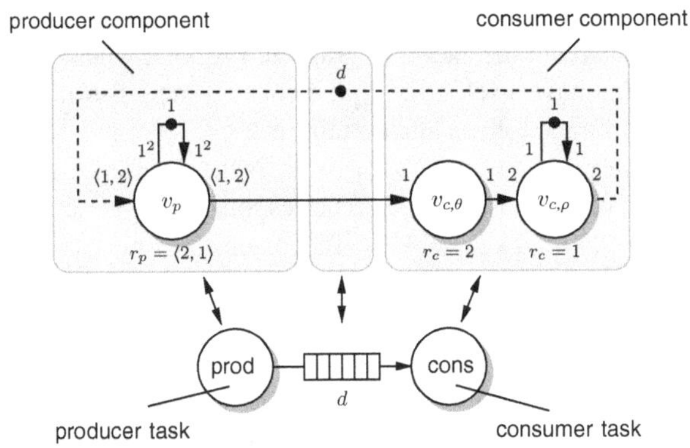

Fig. 6.3 Example buffer capacity problem with a producer and consumer task

that form a partitioning of the graph. This is exemplified in Fig. 6.3 by a producer component that models the producer task, and a consumer component that models consumer task. Note that the dataflow model also includes an edge from the consumer component to the producer component. This is because a task in the implementation only starts when there are sufficient free buffer locations, i.e. space, in all its output buffers. As already exemplified in Fig. 6.2, this type of edges are represented with dashed arrows.

Next, Section 6.1.1 provides a brief introduction to CSDF graphs and the terminology used in the analysis. For further details and examples, see e.g. [10, 15, 44]. There after, in Section 6.1.2, we show how to compute a sufficient buffer capacity d given a requirement on the minimum throughput of the complete dataflow graph.

6.1.1 Cyclo-static Dataflow (CSDF) Graphs

A Cyclo-Static Dataflow (CSDF) graph [24] is a directed graph $G = (V, E, \delta, \tau, \pi, \gamma, \kappa)$ that consists of a finite set of *actors* V, and a set of directed *edges*, $E = \{(v_i, v_j) | v_i, v_j \in V\}$. Actors synchronise by communicating *tokens* over edges. This is exemplified in Fig. 6.3, where tokens, modelling data, flow from actor v_p to the actors $v_{c,\theta}$ and $v_{c,\rho}$, and tokens, modelling space, flow from $v_{c,\theta}$ to v_p. The graph G has an initial token placement $\delta : E \to \mathbb{N}$, as exemplified by the single token on the *self edges* of v_p and $v_{c,\rho}$, and the d tokens on the edge between them.

An actor v_i has $\kappa(v_i)$ distinct *phases* of execution, with $\kappa : V \to \mathbb{N}$, and transitions from phase to phase in a cyclic fashion. An actor is enabled to *fire* when the number of tokens that will be consumed is available on all its input edges. The number of tokens consumed in a firing k by actor v_i is determined by the edge $e = (v_j, v_i)$ and the current phase of the token consuming actor, $\gamma : E \times \mathbb{N} \to \mathbb{N}$, and therefore equals $\gamma(e, ((k-1) \bmod \kappa(v_i)) + 1)$ tokens. The specified number of tokens is consumed *atomically* from all input edges when the actor is started. By introducing a self edge (with tokens), the number of simultaneous firings of an actor is restricted. This is used in the example graph in Fig. 6.3, where actor $v_{c,\theta}$ models *latency*, i.e. it does not have a self edge, and actor $v_{c,\rho}$ models *throughput*, i.e. it can only consume and produce tokens at a certain rate.

The *response time* $\tau(v_i, f)$, $\tau : V \times \mathbb{N} \to \mathbb{R}$, is the difference between the finish and the start time of phase f of actor v_i. The response time of actor v_i in firing k is therefore $\tau(v_i, ((k-1) \bmod \kappa(v_i)) + 1)$. When actor v_i finishes, it atomically produces the specified number of tokens on each output edge $e = (v_i, v_j)$. The number of tokens produced in a phase are denoted by $\pi : E \times \mathbb{N} \to \mathbb{N}$. In the example, v_p has the response time sequence $\tau_p = \langle 2, 1 \rangle$ and $v_{c,\theta}$ has the response time sequence $\tau_{c,\theta} = \langle 2 \rangle$. In this graph, v_p consumes and produces two tokens in the first phase from the edges to and from $v_{c,\theta}$ and $v_{c,\rho}$, respectively. For brevity, the notation x^y denotes a vector of length y in which each element has value x. In the example, this is seen on the self edge of v_p where $1^2 = \langle 1, 1 \rangle$.

For edge $e = (v_i, v_j)$, we define $\Pi(e) = \sum_{f=1}^{\kappa(v_i)} \pi(e, f)$ as the number of tokens produced in one cyclo-static period, and $\Gamma(e) = \sum_{f=1}^{\kappa(v_j)} \gamma(e, f)$ as the number of tokens consumed in one cyclo-static period. We further define the actor topology Ψ as an $|E| \times |V|$ matrix, where

$$\Psi_{mi} = \begin{cases} \Pi(e_m) & \text{if } e_m = (v_i, v_j) \text{ and } v_i \neq v_j \\ -\Gamma(e_m) & \text{if } e_m = (v_j, v_i) \text{ and } v_i \neq v_j \\ \Pi(e_m) - \Gamma(e_m) & \text{if } e_m = (v_i, v_i) \\ 0 & \text{otherwise} \end{cases}$$

If the rank of Ψ is $|V| - 1$, then a connected CSDF graph is said to be *consistent* [24]. For a consistent CSDF graph, there exists a finite (non-empty) schedule that returns the graph to its original token placement. Thus, the implementation it models requires buffers of finite capacity.

We define the vector $\mathbf{s}$ of length $|V|$, for which $\Psi\mathbf{s} = \mathbf{0}$ holds, and which determines the relative firing frequencies of the cyclo-static periods. The repetition vector $\mathbf{q}$ of the CSDF graph determines the relative firing frequencies of the actors and is given by

$$\mathbf{q} = \Lambda \mathbf{s} \quad \text{with} \quad \Lambda_{ik} = \begin{cases} \kappa(v_i) & \text{if } i = k \\ 0 & \text{otherwise} \end{cases}$$

The repetition rate q_i of actor v_i is therefore the number of phases of v_i within one cyclo-static period times the relative firing frequency of the cyclo-static period. For the example in Fig. 6.3, the vector **s** is found to be $[2\,6\,3]^{\mathrm{T}}$, and the repetition vector is $\mathbf{q} = [4\,6\,3]^{\mathrm{T}}$.

For a *strongly connected* and consistent CSDF graph, we specify the required minimum throughput as the requirement that every actor v_i needs to fire q_i times in a period μ [44]. In [130] it is shown how also latency requirements can be taken into account. Given such a requirement on the period μ, sufficient buffer capacities are computed, as discussed in Section 6.1.2.

A CSDF graph is said to execute in a *self-timed* manner when actors start execution as soon as they are enabled. An important property is that self-timed execution of a strongly connected CSDF graph is *monotonic* in time [198]. This means that no decrease in response time or start time of any firing k of any actor v_i can lead to a later enabling of firing l of actor v_j. We return to discuss the importance of this property in the following section.

6.1.2 Buffer Capacity Computation

Two conditions must hold to compute the buffer capacity using dataflow analysis [15]. First, there must be a one-to-one relation between components in the dataflow graph and tasks (e.g. NIs) in the implementation. Second, the model of each component must be *conservative*. That is, if component models task, then it should hold that if data arrives not later at the input of the task than tokens arrive at the input of the component in the model, then data should be produced not later by the task than tokens are produced by component. The mentioned relation between token arrival and production times is required to hold for tokens that represent *data as well as space*. As previously mentioned, the self-timed execution of a strongly connected CSDF graph is monotonic, i.e. there are no scheduling anomalies *in the model*. Together with conservative component models, this means that *bounds on buffer sizes, throughput and latency are valid even if a component produces or consumes faster in the actual implementation*. Thus, in the implementation the scheduling order can change, but the bounds are still valid.

In addition to a model that fulfils the aforementioned conditions, we also need an algorithm to perform the analysis. There are multiple existing algorithms, e.g. [15, 44, 188, 197], and the choice of an algorithm enables a trade-off between the tightness of the bounds and the execution time. In this work, we choose to use a low-complexity (polynomial) approximation technique [15]. In this algorithm, which is described in depth in [15], a schedule is constructed for each actor individually that satisfies the throughput constraint. Subsequently, token production and consumption times resulting from these schedules are linearly bounded. Using these linear bounds, sufficient differences in start times of the individual schedules are derived such that tokens are always produced before they are consumed. These minimal differences in start times form the constraints in a network flow problem that computes minimal start times that satisfy these constraints, thereby minimising the required buffer sizes. Buffer sizes are in the end determined using the computed start times together with the linear bounds on token production and consumption times.

With a conservative model at the component level and one of the analysis algorithms in [15, 44, 188, 197] it is possible to derive sufficient buffer capacities. For example, assuming that the producer and consumer components in Fig. 6.3 conservatively model the producer and consumer task, we can apply the algorithm in [15] to the dataflow graph. A sufficient buffer capacity is computed by finding a token placement d such that the throughput constraint is satisfied, e.g. $\mu = 6$. For the example graph, we have that the graph deadlocks with $d = 2$, while with $d = 3$ the graph has a period of 16. The throughput constraint is satisfied with $d = 7$.

We proceed by showing what is required from a network for these modelling techniques to be applicable. Then we demonstrate how to construct multiple different models of a network channel and evaluate the applicability of the proposed technique for buffer sizing and performance verification.

6.2 Network Requirements

In this section we discuss what must be considered in the NoC architecture to construct a channel model as shown in Fig. 6.2, i.e. of the streaming point-to-point communication. First, the channel *flow control must be blocking*. In other words, data is never lost if a buffer is full, and reading from an empty buffer causes the reading party to stall until data is available. This is the most common way to implement lossless flow control in NoCs, and is used in e.g. [26, 63, 90, 95, 108] (and probably many others), although not in [157].

Second, we require that *all the arbitration points of a channel can be modelled as latency-rate servers* [185], *independent of other channels*. Any number of arbitration points is allowed, and the resources (such as buffers and links) do not have to be statically partitioned to individual channels. Note that the arbitration scheme does not have to be TDM-based and the sharing not composable. In fact, latency-rate characterisation is possible for any starvation-free arbiter. Examples of NoCs that fulfil these requirements are the TDM-based NoCs in [63, 90, 108], and that all have a single arbitration point per channel. Other examples, with multiple arbitration points are [95] and [26] that use round-robin and rate-controlled static-priority arbitration, respectively.

For a NoC architecture that fulfils the requirements, the latency and throughput must be *conservatively modelled*. Although this initially might sound like an easy task, the actual NoC implementation has a wide range of mechanisms (pipelining, arbitration, header insertion, packetisation, etc.) that affect the latency and throughput and have to be taken into account. We exemplify this in the following section where we look at the behaviour of our proposed network.

6.3 Network Behaviour

The latency and throughput of a channel is determined by the resources allocated to a connection, and the direction of the channel, i.e. request or response. As we have seen in Chapter 4, the allocation is captured by the slot table and the path, T_{req}, ϕ_{req}, T_{resp} and ϕ_{resp} for request and response channel, respectively. As suggested in Fig. 6.2, when analysing the request channel, we must also take the response channel into account and vice versa. This is due to the end-to-end flow control that travels back on the channel in the opposite direction. Note that credits travelling on the request channel do not affect data on the request channel (rather they are embedded in the header as shown in Chapter 3), and similar for the response channel.

As we will see in the following sections, the behaviour depends on the number of slots reserved and the distance between them (Section 6.3.1), the number of slots that contain headers (Section 6.3.2), the length of the path through the network (Section 6.3.3), and the availability of flow-control credits (Section 6.3.4). Additionally, a number of architectural constants affect the temporal behaviour. We have already seen the most important ones in Table 3.1. For the purpose of the analysis in this chapter, Table 6.1 adds additional constants that are particular to the temporal behaviour of the NI. We now present the different contributions in the order they appear when a data word traverses a channel and credits are returned.

Table 6.1 Constants particular to the temporal behaviour of the NI

Symbol	Description	Value	Unit
p_n	TDM slot table period	$s_{\text{flit}} \times s_{tbl}$	Cycles
$\theta_{p,NI}$	NI (de)packetisation latency	1	Cycles
$\theta_{d,NI}$	NI data pipelining latency	2	Cycles
$\theta_{c,NI}$	NI credit pipelining latency	2	Cycles

6.3.1 Slot Table Injection

Data injection is regulated by the set of slots reserved on the forward channel, T_d. As discussed in Chapter 3, all slot tables in the network have the same number of slots, s_{tbl} and the flit size is fixed at s_{flit}. Thus, the period in cycles, denoted p_n, is the same throughout the NoC. As we have seen in Chapter 4, the number of slots reserved affects the throughput and latency of the channel whereas the distances between reserved slots only contributes to the latency. In the channel model, we use $|T_d|$ to denote the number of slots in the set of reserved slots (i.e. the size).

Definition 15 Let $d_d(T)$ denote the *upper bound* on the latency, in network cycles, from the arrival of a word at the ingress point of the channel until the average data rate can be sustained when using the slots in T.

Exhaustive search over the possible arrival times is used to determine $d_d(T_d)$.[1] For most practical cases, however, the latter is simply the largest distance between two allocated slots. This is exemplified in Fig. 6.4, by $T_d = \{1, 2, 3, 5, 8, 9\}$ with $|T_d| = 6$ and $d_d(T_d) = 3s_{\text{flit}}$, which happens when a single word appears in the first cycle of slot 5.

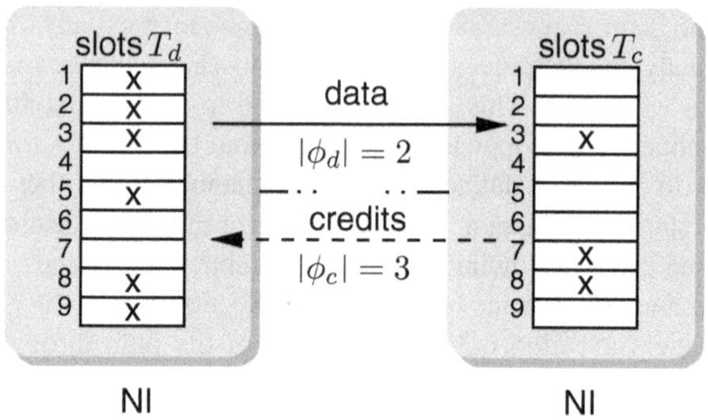

Fig. 6.4 Example channel with slot tables and paths

6.3.2 Header Insertion

Besides data, the forward channel also carries headers. This affects the (net) throughput of the channel, and it is necessary to bound how much of the capacity is used to send headers rather than data. As discussed in Chapter 3, the header insertion is governed by the NI, based on the reservations in the slot table, and the maximum packet size, s_{pkt}. The header has a fixed size of s_{hdr} words, with $s_{\text{hdr}} < s_{\text{flit}}$. For a group of consecutive slots, the first one always has to include the path, and hence a packet header. Consider for example T_d in Fig. 6.4 where slots 5 and 8 include a packet header (slot 1 follows after slot 9). In addition, with $s_{\text{pkt}} = 4$, also slot 3 includes a packet header (as it comes after 8, 9, 1 and 2). We must provide a *lower bound* on the rate of data sent, and hence an upper bound on how many phits are used for headers rather than data.

Definition 16 Let $\hat{h}(T)$ denote an *upper bound* on the number of headers inserted during a period of p_n when using the slots in T.

[1] In fact, it suffices to exhaustively evaluate all intervals that start with a reserved slot and end with an empty slot [142].

In the example, $\hat{h}(T_d) = 3$, which occurs, for example, if we start in slot 4. Note that the bound assumes that data is present for the whole duration of p_n.

6.3.3 Path Latency

As all rate regulation is done in the NI, the traversal of the router network only adds to the latency of the channel, and does not affect the throughput. The latency for a path ϕ depends on the number of hops, denoted $|\phi|$, the pipelining depth of the routers (which is equal to the flit size s_{flit}), and the de-packetisation latency of the NI, denoted $\theta_{p,NI}$.

Definition 17 Let $\theta_p(\phi) = \theta_{p,NI} + |\phi| s_{\text{flit}}$ denote the *path latency* of a path ϕ.

That is, the path latency is determined by the time required to (de)packetise the flit plus the time it takes for the complete flit to traverse the NoC. In Fig. 6.4, $\theta_p(\phi_d) = \theta_{p,NI} + 2s_{\text{flit}}$.

6.3.4 Return of Credits

As discussed in Chapter 3, credit-based end-to-end flow control works as follows. Whenever a word is scheduled for injection in the router network, a counter in the sending NI is decremented. If the counter reaches zero, no more data is sent for the channel in question. When words are consumed from the receiving NI, credits are accumulated and sent back. Hence, there are credits going in the reverse direction, affecting the latency and throughput of the channel.

The credits are returned in the packet headers of the reverse channel and hence do not interfere with the data on the reverse channel. The number of headers depend on the set of slots on the reverse channel T_c. To conservatively model the return of credits, we need to determine a *lower bound* on the rate at which they are sent back, and an *upper bound* on the latency before they are sent. This is to be compared with the bounds determined for the injection of data in Section 6.3.1.

Definition 18 Let $\check{h}(T)$ denote a *lower bound* on the number of headers inserted during any interval p_n when using the slots in T.

For each header, a maximum of s_{crd} credits are sent, as discussed in Chapter 3. In addition to bounds on the rate of credits, we must also bound the latency.

Definition 19 Let $d_c(T)$ denote an *upper bound* on the latency between the arrival of credits (i.e. a word is consumed from the receiving NI) until the average credit rate can be sustained when using the slots in T.

When deriving the latency bound, the slot table allocation, as well as the maximum packet size is taken into account, just as for the data. Looking at the example, $d_c(T_c) = 4s_{\text{flit}}$ and $\check{h}(T_c) = 2$, which happens when starting in slot 9.

The credit return is the last mechanism that affects the latency and throughput of a channel, and we now continue by constructing a model that captures all the different mechanisms.

6.4 Channel Model

In this section we show how to construct a dataflow graph that conservatively models a network channel using the expressions we derived in Section 6.3. We start with a coarse model in Section 6.4.1 and successively refine it until we arrive at our final model in Section 6.4.4. Section 6.4.5 complements the channel model by capturing the behaviour of the protocol shells. In Section 6.5 we use the proposed channel models together with models of the application and apply dataflow analysis [15, 44] to the constructed CSDF graph to determine conservative bounds on the required buffer capacities. Finally, to demonstrate the full potential, the model is used in Chapter 7 to verify application-level performance guarantees, using formal analysis.

6.4.1 Fixed Latency

Our first model, depicted in Fig. 6.5a, has only one actor v_{cd}, with a single token on the self edge. This prohibits an execution to start before the previous execution has finished. As seen in the figure, the actor only fires when buffer space is available in the consumer buffer, denoted β_c, and then frees up space in the producer buffer, denoted β_p. The response time of the actor, τ_{cd}, appearing below the graph, captures the worst-case latency a data word can ever experience. This happens when a word arrives and there are no credits available in the producer NI. Next, we present the four terms that together constitute the response time.

The first term, $\theta_c(T_c) = \theta_{c,NI} + d_c(T_c)$, captures the worst-case latency for the injection of credits. The latency is a sum of (1) the maximum cycles spent updating the credit counter on data consumption and the maximum latency until the credits are seen by the NI scheduler, and (2) the maximum number of cycles without any slots reserved. The second term, $\theta_p(\phi_c)$, corresponds to the time required in the router network to return the credits to the producer NI. With data and credits available, it only remains to bound the time until the data is available in the consumer buffer.

Similar to the injection of credits, $\theta_d(T_d) = \theta_{d,NI} + d_d(T_d)$ bounds the latency experienced by a data word in the sending NI. The latency consist of (1) the number of cycles before a word that is accepted by the NI is seen by the scheduler, and (2) the worst-case latency for data. The fourth and last term is attributable to the router network in the forward direction, which adds a latency of $\theta_p(\phi_d)$.

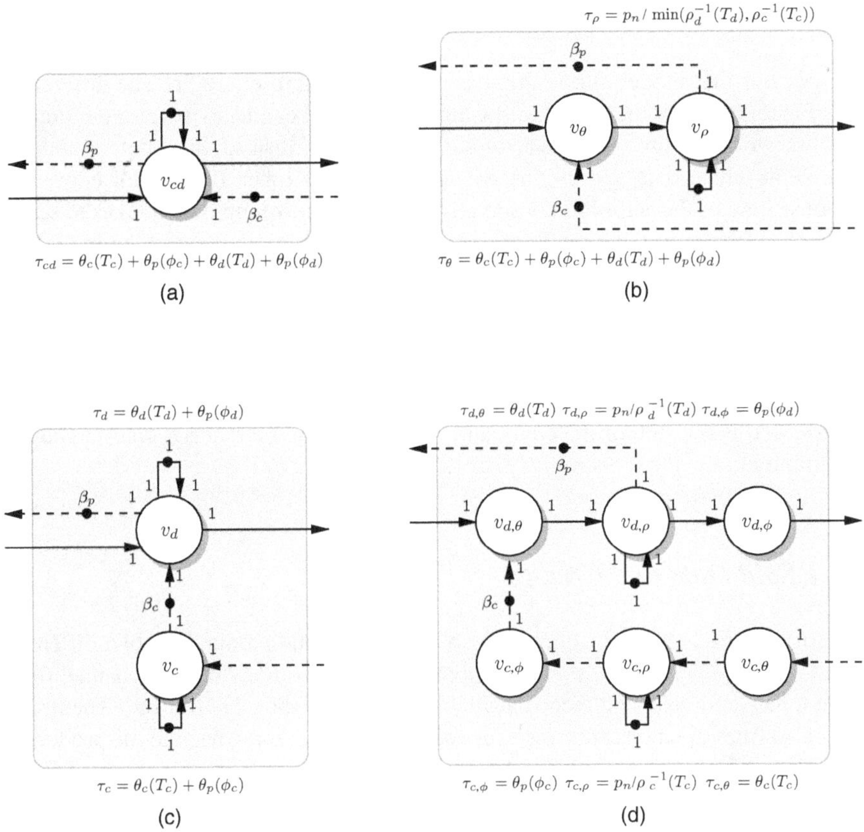

Fig. 6.5 Different channel models. (**a**) Data and credits joined. (**b**) Data and credits joined, latency and rate split. (**c**) Data and credits split. (**d**) Data and credits split, latency and rate split

The model in Fig. 6.5a is sufficient to model the NoC channels and perform buffer sizing and application-level performance analysis. It is, however, overly conservative as it does not distinguish between credits and data, and assumes a worst-case arbiter state for every data and credit item that is sent. Note in particular that only latencies appear in the model. The *number of slots* reserved, for data as well as credits, are not taken into account, which is overly pessimistic.

Next, we show how it is possible to refine the model along two different axes. First, by looking over a larger interval we can create less conservative models. If data/credits arrive fast enough, we only have to assume the worst-case state for the first item. For subsequent items we have more knowledge about the state [185]. This leads to a model where *latency and rate are split*. Second, by distinguishing between the forwarding of data and return of credits, we capture the fact that the two happen in parallel. This leads to a model where *data and credits are split*. Finally, we present a model that combines both these refinements.

6.4.2 Split Latency and Rate

We split our first model into multiple actors according to Fig. 6.5b. The difference with Fig. 6.5a is that the latency, now modelled by v_θ, can be experienced by more than one word at a time, i.e. the actor has no self edge. Instead, the actor v_ρ bounds the rate at which data and credits are sent. With one token on the self edge, the response time of the actor is a conservative bound on how long time it takes to serve one word after an initial latency τ_θ. As seen in the figure, the response time is the period of the TDM wheel, p_n, divided by *the minimum* of the maximum number of data words and maximum number of credits. The number of data words is bounded from above by $\rho_d^{-1}(T_d) = s_{\text{flit}}|T_d| - \hat{h}(T_d)s_{\text{hdr}}$, i.e. the number of words reserved during p_n minus the maximum space required for headers. Similarly, the credits are bounded by $\rho_h^{-1}(T_c) = \check{h}(T_c)s_{\text{crd}}$. In this channel model, we see that the latency experienced is the sum of the credit and data latency, and the rate is determined by the minimum, i.e. the most limiting, of the two.

6.4.3 Split Data and Credits

Our third model, shown in Fig. 6.5c, splits the data and credits into two different actors, v_d and v_c. Actor v_d that models the arbitration on the forward channel fires when it has input data and credits available, as seen by the edge from v_c. The firing of v_d also frees up one buffer space, as seen by the edge going back to the producer. The return of credits is modelled by v_c that fires when a word is consumed from β_c. The response times of the actors, appearing above and below the graph, capture the time it takes for a word/credit to be scheduled by the NI and traverse the path through the network. Compared to our first model, we see that the latency for data and credits now appear as the response time of v_d and v_c respectively. We also see the asymmetry between the producer buffer β_p that is local, and the consumer buffer β_c that is located in the receiving NI.

6.4.4 Final Model

By splitting the model into a data and credit part, as well as a latency and rate part, we arrive at our final model, shown in Fig. 6.5d. In the forward direction, data experiences scheduling latency and rate regulation in the NI, modelled by $v_{d,\theta}$ and $v_{d,\rho}$. The router network also adds latency, $v_{d,\phi}$. For the return of credits, the situation is similar. The NI is modelled by $v_{c,\theta}$ and $v_{c,\rho}$, and the router network by $v_{c,\phi}$. In our final model, we use independent actors to model latency/throughput and data/credits.

Note that the channel models are *independent of the application* using it. If the application behaviour changes, or the model of an application is refined, nothing has to be modified in the channel model. Similarly, the application model is not affected if the channel model is replaced or refined, i.e. by choosing one of the aforementioned levels of detail.

6.4.5 Shell Model

So far we have only modelled the NI and router network (including any pipelined links), and to apply the model to point-to-point memory-mapped communication we must also model the protocol shells. As discussed in Chapter 3, the shell is closely coupled to the IP and the protocol it uses. In Table 6.2 we summarise the behaviour of the shell as described in Chapter 3, and show how the current shell implementations transform the read and write messages of a memory-mapped protocol into a number of words of streaming data. As seen in the table, the size of a burst, from the perspective of the NI, depends on the burst size of the IP, the transaction type, the channel direction, and the size of the message headers. Table 6.2 shows the case when an element of the memory-mapped protocol fits within the width of the network links. As seen in the table, a read operation requires the read command (plus potential flags) and the address to be sent as a request. The response, in turn, carries the actual data as well as status information. For a write operation, the command, flags and address are sent together with the data (and potentially a write mask). When executing a write operation, the target may also return status information and error codes.

Table 6.2 Burst sizes for different channel types

Transaction	Direction	IP (words)	Shell (words)	Total (words)
Read	Request	0	$h_{\mathrm{req},r}$	$h_{\mathrm{req},r}$
Read	Response	b	$h_{\mathrm{resp},r}$	$b + h_{\mathrm{resp},r}$
Write	Request	b	$h_{\mathrm{req},w}$	$b + h_{\mathrm{req},w}$
Write	Response	0	$h_{\mathrm{resp},w}$	$h_{\mathrm{resp},w}$

Note that the shell model only captures command handshakes and data transfers for request and responses, i.e. point-to-point communication. Higher-level protocol issues, e.g. coupling between requests and response or dependencies between different request from the same initiator are captured in the model of the application together with the behaviour of the initiator and target buses, as discussed in Chapter 7.

6.5 Buffer Sizing

In this section we demonstrate the applicability of the model by comparing the run time and buffer sizes derived using our approach with those of [39, 56]. Using the terminology of [39], we will hereafter refer to the two approaches as *analytical* and *simulated*, respectively. Moreover, using our proposed channel model, we also show the differences using the dataflow model with fast approximation algorithms [15] and exhaustive back-tracking [44]. Thus, we use the same model as input, but two different tools in the analysis. The run time is measured by using the Linux command *time*, looking at the user time, including potential child processes. The reason we use this command is that the dataflow analysis tools are separate binaries that are called for every connection that is analysed in the NoC design flow [62]. For the dataflow analysis, the time includes the forking of the parent process, the writing of XML files describing the CSDF graphs to disk, and then reading and parsing of those files.

The first step to comparing with [39, 56] is to adopt an equivalent application model. There after, we apply the methodologies to a range of synthetic benchmarks, a mobile phone SoC design, and a set-top box SoC.

6.5.1 Modelling the Application

To facilitate comparison with existing work, we choose to employ a model that subsumes [39, 56], where the input specification is done per connection, and contains a transaction type, i.e. read or write, a transaction size $b \in \mathbb{N}$ in words, and a period $p \in \mathbb{N}$ in *network clock cycles* (if the IP and NoC are using different clock frequencies).

Similar to [39], our channel model is based on the notion of streaming point-to-point communication between a producer and consumer. To determine the sizes of the four connection buffers,[2] where the initiator and target acts as both producer and consumer, we first look at the data going from initiator to target to determine $\beta_{\text{req},i}$ and $\beta_{\text{req},t}$. Thereafter, we swap the roles and let the target be the producer and the initiator the consumer to determine $\beta_{\text{resp},t}$ and $\beta_{\text{resp},i}$.

Dividing the buffer calculation for request and response channel into two separate steps, implicitly assumes that the production and consumption inside the IPs is completely decoupled, i.e. there is no relation between consumption and production times of the IP. Again, this is to keep the model comparable to [39, 56], and is addressed in the case study in Chapter 7.

Figure 6.6a shows the models we adopt for periodic producers. Since we model an IP that produces words on a bus interface, we know that a burst of more than one word cannot be produced in one cycle. This is reflected in the model, where only one token is produced per actor phase. Space is acquired in the first b phases, each taking one cycle. Data is released in the last b phases, also taking one cycle each. Both actors have a cumulative response time of p. Note that the model of the consumer is completely symmetrical, with the only difference being which edges represent data and credits. Hence, the consumer acquired data in the first b phases, and releases the space the data occupied in the last b phases.

---

[2] Two for the request channel and two for the response channel.

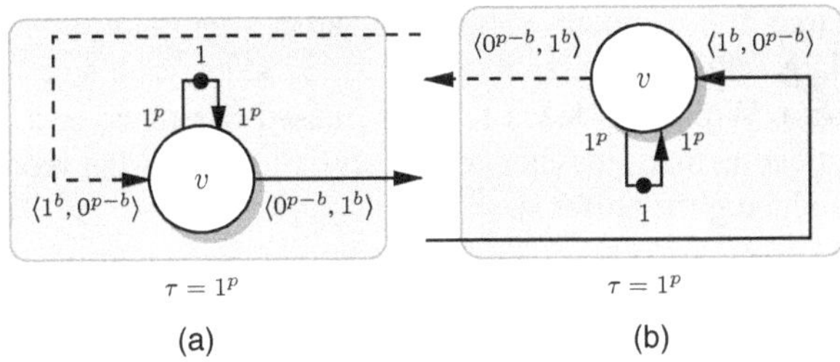

Fig. 6.6 IP models used for buffer-size comparison. (**a**) Producer. (**b**) Consumer

If the producer acquires its space later in the implementation, then the buffer capacities computed with this dataflow model are still sufficient. This is because space will arrive in time to allow for consumption times according to the model, which implies that space arrives in time for these later consumption times. If the producer actually releases its data earlier, then still the computed buffer capacities are sufficient. This is because an earlier production of data can only lead to an earlier enabling of the data consumer, i.e. the NI. This is because of the one-to-one relation between components in the implementation and the model, together with the fact that the dataflow model is monotonic in the start and response times. The reasoning for the case in which the IP consumes instead of produces data is symmetric.

6.5.2 Synthetic Benchmarks

To assess the performance over a broad range of designs we choose to compare the different algorithms on a set of randomly generated use-cases. The benchmarks follow the communication patterns of real SoCs, with bottleneck communication, characterising designs with shared off-chip memory, involving a few cores in most communication. All generated designs have 40 cores, with an initiator and a target port, connected by 100 connections. Throughput and latency requirements are varied across four bins respectively. This reflects for example a video SoC where video flows have high throughput requirements, audio has low throughput needs, and the control flows have low throughput needs but are latency critical. A total of 1,000 benchmarks are evaluated, *using the proposed channel model more than 200,000 times in total, with widely varying requirements.*

Figure 6.7a shows the distribution of the total buffering requirements, relative to the analytical approach [56]. As seen in the figure, both the simulation-based algorithm and the algorithms using our dataflow model result in significant reductions on the total buffer size. Averaging over all the benchmarks, we see a reduction of 36% using the dataflow approximation algorithm, 41% using the exhaustive simulation and 44% when applying the exact dataflow analysis. Moreover, the distribution of relative improvement is much wider for the simulation-based algorithm, ranging all the way from 5% up to 45%. The dataflow model, on the other hand, consistently results in an improvement of more than 30%, and even 35% in the case of exact analysis. The large improvements stem from rather small slot tables (<32 slots), and thereby a large allocation granularity (with 32 slots, each slot corresponds to roughly 63 Mbps). While this leads to an increased burstiness and larger buffers using the analytical method, the dataflow analysis leverages the reduced response times, thereby reducing the buffer sizes.

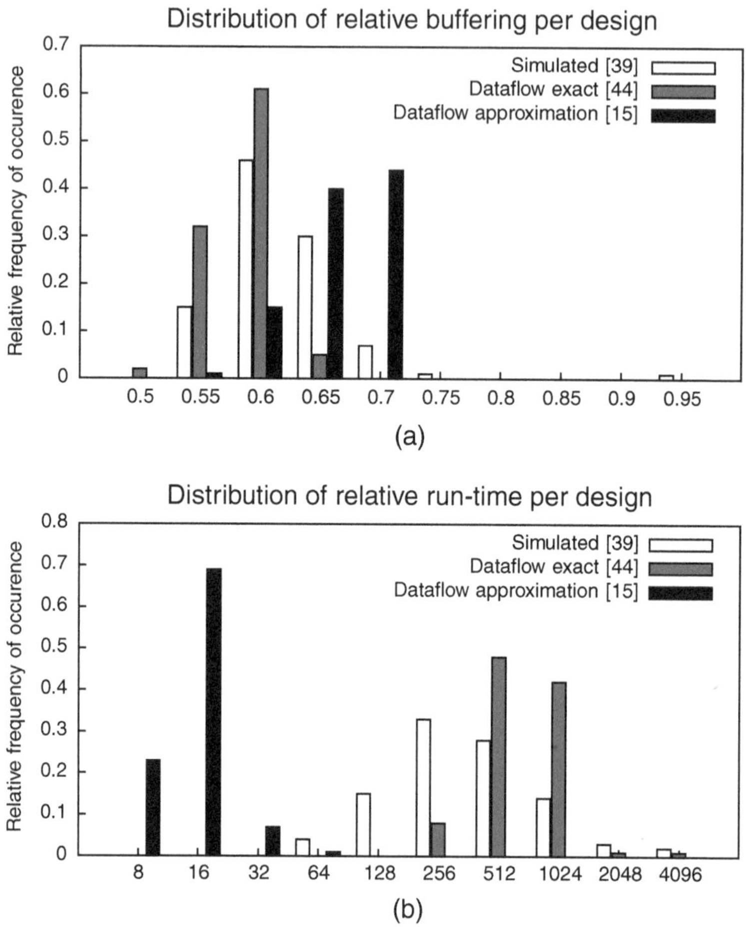

Fig. 6.7 Relative run time (**b**) and buffer size (**a**) compared to [56]

The run times measured when deriving the aforementioned buffer capacities are also compared to [56] and the relative distribution is shown in Fig. 6.7b. It is clear from the experiment that the dataflow approximation algorithm is roughly one order of magnitude slower than the analytical approach, being on average 11 times slower. Note though that the run time is still below a second for an entire SoC design. The exact dataflow analysis and the simulation, on the other hand, are three orders of magnitude slower, averaging at 450 and 520 times the execution time of [56] (but faster algorithms for the exact dataflow analysis exist [188]). The run time of the simulation-based algorithm depends heavily on the period of the producer, network and consumer. As the generated designs use relatively small slot tables, the run times are in the order of minutes.

6.5.3 Mobile Phone SoC

A phone SoC with telecom, storage, audio/video decoding, camera image encoding, image preview and 3D gaming constitutes our first design example. The system has 13 cores (27 ports distributed across an ARM, a TriMedia, two DSPs, a rendering engine, etc.), one off-chip DDR memory, one on-chip SRAM plus a number of peripherals. Communication is done via memory, and the NoC is running at a frequency of 235 MHz.

The total buffer size and the time needed to derive all buffers, for all the use-cases, are shown in Table 6.3. As explained in [39], the buffer sizes are determined per use-case and then the maximum for every buffer is used in the architecture. Again, we see that the dataflow-based methods results in improvements of 30 and 34% respectively. The simulation, however, only reduces the buffers by 12%, thus performing significantly worse than for the synthetic benchmarks. Moreover, while the dataflow approximation technique results in run times that are comparable to those seen in the synthetic benchmarks, the exact analysis and the simulation-based approach are roughly 10,000 and 100,000 times slower than [56]. The reason for the increased run time is a large number of low throughput connections with long periods.

Table 6.3 Buffer sizes for mobile phone system

Algorithm	Run time (s)	Total buffers (words)	Impr. (%)
Analytical [56]	0.05	1,025	ref
Simulated [39]	6,845	799	12
Dataflow approx. [15]	0.78	721	30
Dataflow exact [44]	547	680	34

6.5.4 Set-Top Box SoC

Our second design example is a set-top box with four different use-cases, all having hot spots around three SDRAM ports and 100–250 connections. These connections deliver a total throughput of 1–2 Gbps to 75 ports distributed across 25 IP modules. With more than 500 connections to be analysed, this constitutes the largest example. The slot table size is also larger, 67 slots, to accommodate a wide variety of throughput requirements.

As with the mobile phone SoC, we look at the total buffer capacity required across the use-cases and the run time needed to compute the buffer sizes. The results are presented in Table 6.4, except for the simulation-based that had not even finished the first use-case after running 24 h. Simulating one least common multiple for every possible scheduling interleaving is therefore often impractical for a design of this size. For the dataflow analysis, the approximation algorithm is almost as fast as [56], since the ratio of computation and file I/O is far larger than for the previous examples. It should be noted that this time also includes four invocations of the exact algorithm, as the heuristic failed to find a solution. Exclusive use of the exact algorithm, on the other hand, is considerably slower, again due to a very large solution space. The slot table size for this design is also fairly large, leading to smaller discretisation effects, and this benefits the analytical algorithm that uses less conservative bounds. At the same time, the dataflow analysis does not have the opportunity to reduce the buffering requirements by exploiting lower response times, and we see a relative improvement of only 22% for the exact analysis.

Table 6.4 Buffer sizes for set-top box system

Algorithm	Run time (s)	Total buffers (words)	Impr. (%)
Analytical [56]	6.10	9,190	ref
Simulated [39]	–	–	–
Dataflow approx. [15]	6.87	7,818	15
Dataflow exact [44]	15,229	7,170	22

While 7,170 words of buffering in the NIs might seem costly, it should be noted that most of this buffering is located close to the memory controller, and its three ports. It is thus possible to use a few large dual-ported SRAMs rather than dedicated FIFOs. Thereby, the roughly 24 kb worth of buffering occupies only 0.2 mm^2 [149] in a 65-nm CMOS technology.

6.6 Conclusions

In this chapter we have presented the verification part of the design flow. We have shown a number of different possible verification methodologies, and demonstrated in detail how to derive a formal model of a network channel to enable buffer sizing and application-level performance analysis. We summarise by evaluating the contribution to the different high-level requirements in Chapter 1.

The verification flow contributes to the *scalability* of the interconnect by enabling performance evaluation per application, thanks to the composability of the interconnect. This is an important difference with existing platforms where the size of the system negatively impacts simulation speed [166] (more to simulate and more behaviours to cover), making system-level simulation untenable [152] as a verification technique.

The contribution to *diversity* lies in the many different ways the verification can be performed, i.e. by simulation on multiple levels, or by using formal models of the platform. The latter can also be used both for worst-case and typical-case analysis depending on the requirements of the application. Thus, depending on the requirements and assumptions placed on the application, different types of analysis are offered.

As we have seen in this chapter, the verification flow relies on *composability* to offer scalability. It does, however, not offer any contributions to composability itself. *Predictability*, on the other hand, is the focus of this chapter. We show how to construct a dataflow model of a network channel, requiring only that the specific network uses blocking flow control and that the arbitration points can be modelled as latency-rate servers. We demonstrate that a channel in our network is a latency-rate server, and present a detailed model of such a channel. The proposed model has been evaluated quantitatively by comparing with existing buffer-sizing approaches on a range of SoC designs. If the application has a dataflow model, it can be combined with the presented channel model to include the temporal behaviour of the communication between tasks. This enables verification of firm-real time end-to-end constraints.

Similar to composability, the verification flow has no contribution to the *reconfigurability* of the interconnect. Looking at *automation*, however, there are number of important contributions. The verification flow offers fully automated simulation and evaluation of the interconnect, using either the SystemC, RTL or netlist instantiation or the hardware. Furthermore, after a completed simulation, the flow presents the user with detailed traces as well as high-level performance overviews (e.g. latency, throughput and buffer usage). In addition, the verification flow generates dataflow models of the network channels. Using these models (and possibly models of the applications), buffer sizes are automatically computed and back-annotated.

Chapter 7
FPGA Case Study

Using our example system from Chapter 1, we have seen how the interconnect is dimensioned, how resources are allocated and how the resulting hardware and software is instantiated and verified. In this chapter, we take the last step and demonstrate the diversity, composability, predictability, reconfigurability and automation of our interconnect by creating an actual system instance.[1] This instance, depicted in Fig. 7.1, is equivalent to the system in Fig. 1.4 in that it comprises: a host, three processor cores, an SRAM controller, an audio ADC/DAC, a memory-mapped video subsystem and a peripheral tile with a character display, push buttons and a touch screen controller. However, we now use three homogeneous VLIW processors instead of a μBlaze, an ARM and a VLIW. Moreover, as we shall see, all IPs other than the processors, i.e. the memory and I/O functionality, are integrated on the board level, with on-chip wrappers.

We map the six applications from Fig. 1.5 on our system instance. However, we restrict our evaluation to three applications, namely the initialisation application, the M-JPEG decoder and an the audio post-processing filter. The reason we choose to focus on these three applications is that they have diverse behaviours and real-time requirements, and use both streaming and memory-mapped communication. The audio filter represents one extreme, with firm real-time requirements. To verify the requirements, the filter must be implemented in such a way that a conservative schedule for the entire application can be derived at design time, thus guaranteeing periodicity of the ADC and DAC. Such restrictions are not desirable, and also not necessary for the decoder that has a data-dependent behaviour, which is typical for compression and decompression functions [107]. Thus, the amount of data produced or consumed, and the processing delay varies over time. Despite the dynamic temporal behaviour, however, we want a tight estimated on the decoding time for a given set of test images. Lastly, for the initialisation there are no requirements on the temporal behaviour. In addition to the diversity in behaviours and requirements, the selected applications make use of both memory-mapped and streaming communication, and share both the network and the initiator bus of the SRAM.

[1] In this chapter we do not demonstrate scalability due to the limited resources on our target FPGA.

A. Hansson, K. Goossens, *On-Chip Interconnect with Aelite*, Embedded Systems,
DOI 10.1007/978-1-4419-6865-4_7, © Springer Science+Business Media, LLC 2011

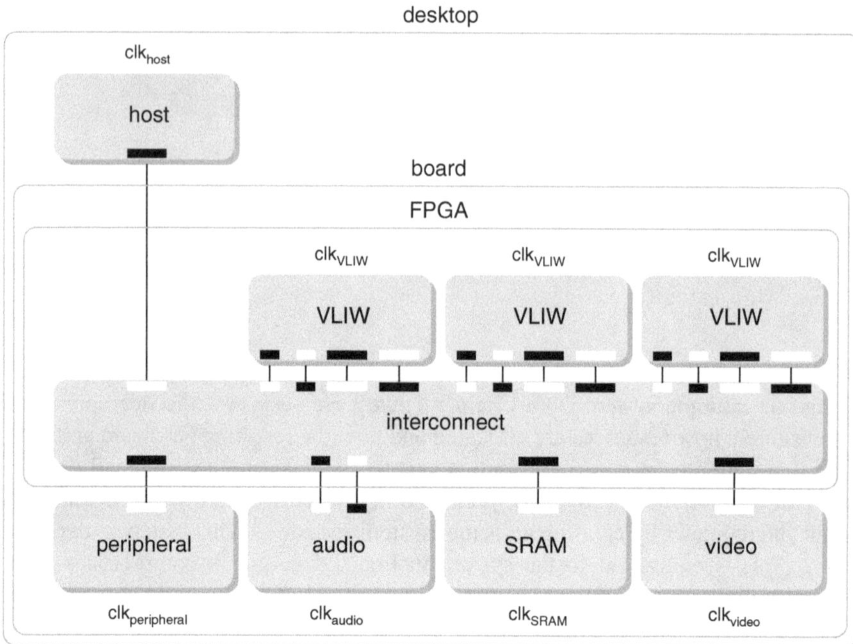

Fig. 7.1 Example platform instance

We start by introducing the hardware platform (Section 7.1) and software platform (Section 7.2). Then, the applications are mapped to our instance (Section 7.3), and we show how composability enables independent analysis of the applications, using different techniques (Section 7.4). We end this chapter with conclusions (Section 7.5)

7.1 Hardware Platform

For our system instance, we use an RC340 board from Agility DS. We place the three processors, the interconnect and the display controller on the Xilinx Virtex4 LX-160 FPGA. On the FPGA, we also include wrappers for all the board-level interfaces. At the board level, the FPGA is connected to a USB 2.0 interface through which the host interfaces with the rest of the system. Also at the board level, our system contains a character LCD display, a touch screen controller, an audio codec, a DVI transmitter and two banks of 8 MB SRAM. One memory bank is used by the SRAM controller, and one by the display controller. Various wrappers bridge between board-specific APIs used by Agility and the industry-standard DTL [49] protocol used by the VLIW. Currently, streaming interfaces are only used by the audio codec, whereas the other wrappers all use memory-mapped interfaces.

The audio codec, SRAM controller and video memory are introduced in Chapter 1. The peripheral tile, shown in Fig. 7.2, internally contains a target bus that is responsible for multiplexing between the character display, touch-screen controller, buttons and timers based on the address. Note the reuse of the bus with a static address decoder from Chapter 3. Our platform instance has only one shared memory. While this is common, either for cost reasons or due to a limited number of pins [107], a single shared memory is inherently non-scalable, as the performance is directly affected by the amount of applications sharing it [90] (see the discussion on architectural scalability in Chapter 2). It should be noted that the interconnect does not constrain the number of memories, but currently only supports composable and predictable sharing of SRAMs, as discussed in Chapter 3. Next, we elaborate on the functionality of the two components that are specific for this instance, namely the host and processor tiles.

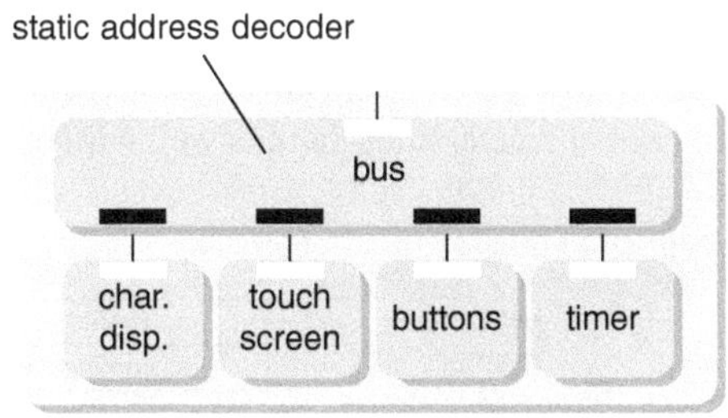

Fig. 7.2 Peripheral tile

7.1.1 Host Tile

Rather than having the host embedded on the FPGA, e.g. in the form of an ARM processor, we choose to let the host interface with the board using a USB connection. Through a board-specific API, it is thereby possible for a normal PC to take the role of the host. By means of the on-chip wrapper, the PC appears as a memory-mapped initiator (with a DTL interface) to the interconnect. Coupled with software libraries for the PC, reads and writes are done through a high-level API, disclosing nothing about the link-level protocol used for the USB communication.

The USB communication libraries together with the run-time API of the interconnect, as described in Chapter 5, enable the PC to open and close connections. The two libraries are compiled into a host binary using any native compiler, such as gcc. The resource allocations themselves are stored in the non-volatile memory of the PC (e.g. a hard disk or flash memory). Thus, in contrast to the control infrastructure introduced in Chapter 3, there is no need for a non-volatile memory on the local control bus of the host. The major drawback with performing all reconfiguration from an external host is that no temporal bounds can be given on the control operations (both due to the USB connection and the PC). It is thus not possible to guarantee upper bounds on the time required to start and stop applications, as done in Chapter 5.

Having the host placed outside the FPGA greatly improves observability and controllability, and makes it easy to control not only the interconnect, but also the other IPs, and then in particular the processors. The binaries of the processors are cross-compiled, as later discussed in Section 7.2, and stored in the non-volatile memory of the PC. Using the run-time libraries of the processors, the PC uploads binaries and data to the processors, and starts and stops execution dynamically at run time. All the aforementioned operations require access to a target port on the processors, and the host uses the connections of the initialisation application to reach those ports. We now describe the processors in greater depth.

7.1.2 Processor Tiles

The processor tiles, depicted in Fig. 7.3, are based on Silicon Hive [175] VLIW processing cores. The processor are customisable, making it possible to adapt the costs and the performances to a given application, as advocated in [92, 166]. The architecture has a number of characteristics that fits well with our requirements from Chapter 1.

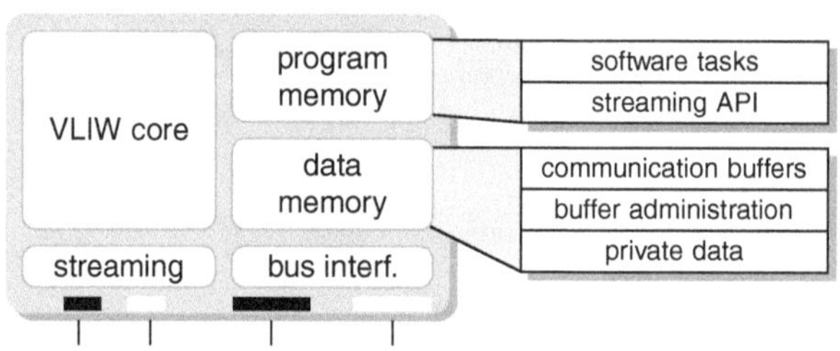

Fig. 7.3 Processor tile

The processors contribute to the architectural *scalability* by having their own private program and data memory, in our instance 32 kbyte program memory and 32 kbyte data memory. Having memory distributed over the different processing devices provides higher throughput with lower latency, which results in a higher performance at a lower power consumption [180]. Instructions are executed from a local memory thus removing the latency-critical reads from external memories. Moreover, the memories are accessible through a target port on the processors bus interfaces, thus avoiding central bottlenecks by forming a distributed memory together with the dedicated memory tile. Arbitration between multiple external initiator ports is implemented by a (predictable) round-robin arbiter.

The Silicon Hive processors also contribute to the *diversity* by offering both memory-mapped and streaming communication interfaces. The cores use a memory-mapped architecture, with support for multiple load-store units. The initiator interface on the processor enables reads and writes to memories external to the processor. In addition to the traditional load-store units, the processor template also has send-receive units that act as instruction-mapped streaming interfaces [107, 191]. For applications that use streaming communication, e.g. our audio filter, it is thus possible to match the architecture to the application, as advocated in [92].

The processor template is also appropriate for applications that require *predictability*. The processor core has no caches that complicate performance analysis. Furthermore, unlike a super-scalar processor, the VLIW has no bypassing or hazard detection mechanisms. As we will see in Section 7.4, the aforementioned properties enable us to easily determine the execution time of tasks by analysis of the instruction schedule of the VLIW.

The major limitation of the Silicon Hive processors is that they do not support pre-emptive multi-tasking. It is hence not possible to share a processor in a composable way, i.e. eliminating all application interference. As a result, we do not allow multiplexing of task belonging to *different applications* on the same processor. Note, however, that this is not a fundamental issue, and merely a result of the specific processor architecture. Moreover, we allow processor sharing between tasks belonging to the same application, e.g. by a static-order scheduling strategy, where the order in which the tasks execute are determined before the application is started [14].

7.2 Software Platform

While the hardware of the system plays an important role, there are also two essential pieces of software. First, the *application middleware*, i.e. libraries that facilitate inter-processor communication, further discussed in Section 7.2.1. Second, and most importantly, the *top-level design flow* that helps in defining, realising and verifying the entire system. This software component is discussed in Section 7.2.2.

7.2.1 Application Middleware

To facilitate synchronisation and communication between tasks running on the processors (in our case the VLD, IDCT and CC tasks of the decoder), an implementation of the C-HEAP [144] protocol[2] is offered as a part of the platform. The protocol specifies an API for *token-based FIFO communication*, built on shared memory. Synchronisation is done using pointers in a memory-mapped FIFO-administration structure.

C-HEAP fits well with the proposed interconnect for several reasons. First, it does not use interrupts or target locking for synchronisation, but instead relies on polling. With interrupt-based synchronisation, it is difficult or even impossible to bound the frequency of interrupts and the execution time incurred by them [99]. Second, the local memories of the processors are suitable for mapping communi-

[2] C-HEAP in its entirety is not only a protocol for cooperation and communication between tasks, but also a top-down design methodology and an architectural template [144].

cation buffers, as shown in Fig. 7.3. Using the local memories of the processors enables low-latency access to data with random access in acquired data and space. Moreover, changes in element size and FIFO length are possible even after final silicon realisation. Third, by keeping two copies of the buffer administration, both at the producer and consumer side [144], *all read operations (polling) are local* and *only posted write operations traverse the interconnect* (also known as the *information push paradigm* [99]). The advantage of only writing and never reading remotely is the reduced impact of the interconnect latency. We exemplify the use of the C-HEAP implementation in Section 7.4.

7.2.2 Design Flow

The hardware flow, depicted in Fig. 7.4, takes its starting point in high-level descriptions of the processors, peripherals and their interfaces. The Silicon Hive tooling instantiates the processors used in the design according to their individual descriptions. More details about this process are found in [175]. The tools are configured to generat RTL tailored specifically for FPGA implementation. Using board-specific APIs, the peripherals and their wrappers are instantiated using Agility's tools. The latter directly generate a netlist for the peripherals. Lastly, the IP interface description is used to generate an interconnect instance, all according to the interconnect design flow proposed in this work.

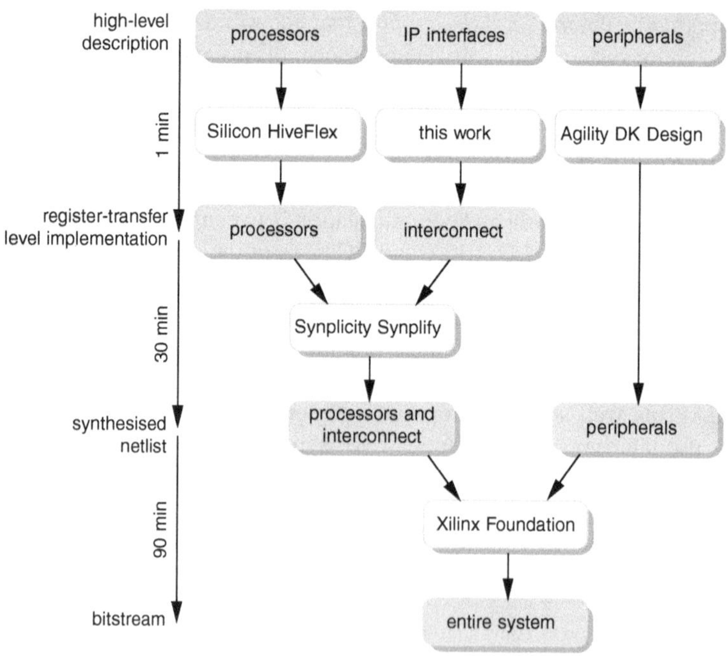

Fig. 7.4 Hardware design flow

The starting point of the software flow, depicted in Fig. 7.5, is the same description of the cores that is used in Fig. 7.4. Here, it is used to generate processor-specific libraries for the retargeteable assembler and linker. These are then used to compile the source code that is to be run on the processors. The end result is microcode, then can be uploaded and executed on the embedded processors. In parallel with the embedded microcode, the code needed to orchestrate the execution on the processors and the configuration of the interconnect is produced and linked together with the Silicon Hive and interconnect run-time libraries, as discussed in Chapter 5.

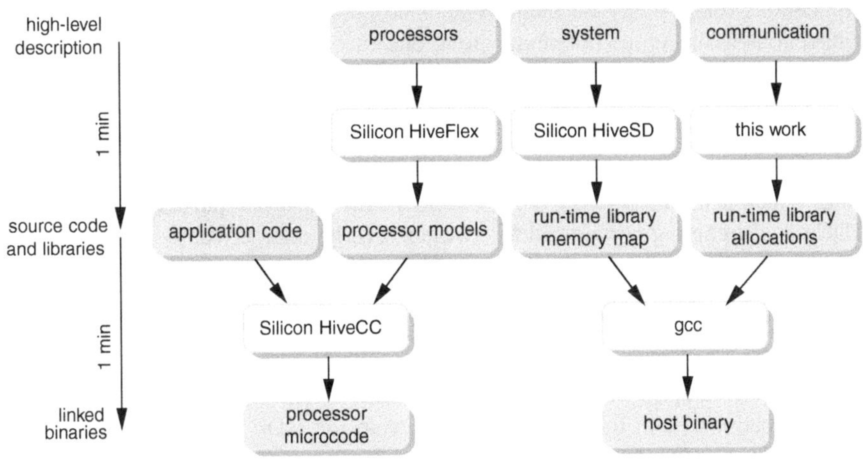

Fig. 7.5 Software design flow

As shown by the time line on the left hand side in Fig. 7.4, the trajectory from high-level specifications to an entire multi-processor system (that fulfils those requirements) takes only a couple of hours. Moreover, as we have seen in Chapter 4, it is possible to change the resource allocation (at compile time rather than design time) after the hardware is fixed. Consequently, it is possible to accommodate new or modified applications, assuming that the requirements do not exceed the resources that are available. Modifications to the host software and application code are performed within minutes, as illustrated on the left hand side of Fig. 7.5. We now show how to apply the flows to our example applications.

7.3 Application Mapping

The starting point for the M-JPEG decoder is sequential C code that is split into multiple tasks according to Fig. 1.5c, i.e. one processor executes the VLD and a second processor is responsible for the IDCT and CC.[3] Communication between

[3] In fact, the parallelisation is done by master students as part of a MSc course where the proposed case study, including the interconnect, is used as part of the experimental platform [76].

the processors is implemented using the C-HEAP API, with FIFO data mapped to circular buffers in the local memory of the receiving processor. The original code is easily ported to the platform (although not optimised), requiring only an explicit mapping of variables and communication buffers to memories. The VLD together with the inverse zig-zag and quantisation occupies 24.3 kB program memory and 6.7 kbyte data memory (not counting the C-HEAP FIFOs). The second processor involved in the M-JPEG decoding executes the IDCT and CC, requiring 13.4 kbyte of program memory. The encoded input image is read from the SRAM by the first processor, and written to the frame buffer by the second processor. The decoder applications relies solely on memory-mapped communication, with four network connections realising the communication: one each for accesses to the SRAM and video, and two (write-only connections) for the inter-processor communication.

The audio post-processing application, shown in Fig. 7.7a, comprises three tasks. First, the source ADC, periodically producing signed 16-bit pulse code modulated stereo samples. Second, the actual filter task, executed on the processor. Third, the DAC, which acts as a periodic sink. The filter task receives input samples from the ADC via a streaming port and adds a two-tap reverberation effect by mixing in past samples that are read from the SRAM. The output is then sent both to the DAC using streaming communication and stored in the background memory for future mixing with the input. Thus, the processor uses streaming for the communication with the ADC and DAC, and shared-memory communication for reading and writing reference samples in the background memory. The filter application requires two connections (with the channels to and from the ADC and DAC sharing one connection), and has one out of three time slots allocated in the target bus of the SRAM.

The last application is the initialisation done by the host. Besides loading encoded JPEG images into the background memory and configuring the peripherals (in this case the sampling rate for the ADC and DAC), the initialisation application is responsible for boot strapping the processors by loading their local program and data memories. Rather than only having two connections, as shown in Fig. 1.5f, this application therefore has five connections, as detailed in Appendix A.

Network resources are allocated to the six different applications using the compile-time flow described in Chapter 4. The amount of resource needed for the filter application is determined based on its firm real-time requirements and strictly periodic behaviour (as described in Section 7.4). For the decoder, the amount of resources requested are based on rough estimates of how much data needs to be communicated *on average* between the different tasks. The performance of the decoder can thus be improved by asking for additional resources, a trade-off left for the application designer. For the initialisation application, we do not specify any latency requirement and request only a minuscule throughput. Furthermore, since we know that the host performs the various initialisation tasks in a serial fashion, we specify two channel trees, for the request and response channels of the initialisation application. While conserving the interconnect resources, as a result, we must also open and close these connections when switching between the different targets, e.g. the peripheral, SRAM or the local memories in one of the processor tiles.

The complete architecture is mapped to the target FPGA, using Synplicity Synplify 8.8 and Xilinx Foundation 9.1, as shown in Fig. 7.4. The resulting design occupies 96 (33%) block RAMs, 25 (26%) DSPs, 18397 (13%) flip flops and 57682 (42%) LUTs. The obtained maximum clock frequency (with the interconnect and processors in one synchronous island) is 48 MHz, and the tools estimate a total equivalent gate count of roughly 7.8 Mgates. As already discussed, the SRAM, the audio codec, the USB μcontroller and the display controller are all integrated on the board level, with only the wrappers occupying resources on the FPGA.

7.4 Performance Verification

We proceed to analyse the performance of the decoder and the filter (the initialisation application has no real-time requirements). The analysis is done independently for the two applications, first looking at the decoder, and then at the filter. Moreover, as we shall see, the analysis technique is different for the two applications.

7.4.1 Soft Real-Time

The decoder is data dependent, in that the different decoding steps require a varying amount of computation (and communication) for different images. An image with a large amount of detail contains more high-frequency components and has a larger file size. As a consequence, more data must be read from the background memory, and it also takes longer to execute the VLD. Therefore, two images, denoted *small* and *large*, are used to evaluate the performance of the decoder. Both images have XGA resolution, and only differ in the amount of details in the picture.

The performance of the decoder is analysed by simulation and by instrumentation on the actual FPGA implementation. For the specific input traces, we evaluate the end-to-end performance of the decoder by using the channel models from Chapter 6 to determine the communication behaviour (in contrast to plain behavioural SystemC models). The results of the evaluation are shown in Fig. 7.6.

Comparing the ideal interconnect with instantaneous communication to the actual FPGA instance in Fig. 7.6, we see that communication has a relatively small contribution to the total decoding time. The difference in time between the two is merely 11% and 2%, for the large and small file size, respectively.[4] Continuing by looking at the simulation results, with the most detailed channel model, the end-to-end performance is within 8% and 2% of the time measured on the FPGA for the two files, respectively. This clearly shows that it is reasonable to use the channel model proposed in Chapter 6 to evaluate the end-to-end performance of soft real-time applications.

---

[4] It should be noted that the processor core we use is extremely simple not optimised in any way for JPEG decoding.

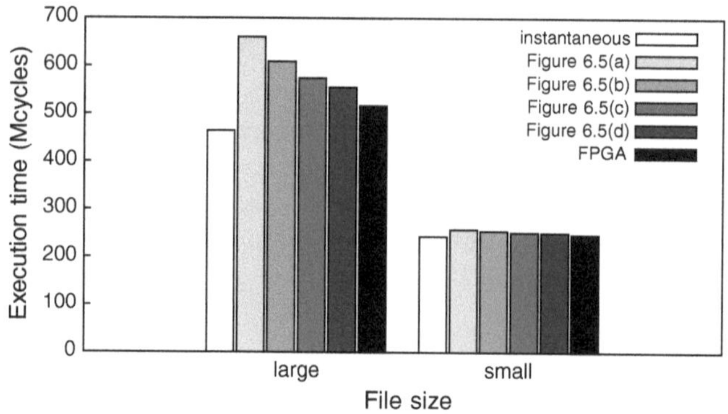

Fig. 7.6 Execution time using different channel models

7.4.2 Firm Real-Time

In this section, we demonstrate how to model the complete filter application as a CSDF graph, and thus enable verification of the firm requirements on end-to-end temporal behaviour. The worst-case analysis of the complete application is made possible by the predictability of the interconnect and the IPs (in particular the processor on which the filter task is executed). We choose to model the filter applications and its mapping to the architecture as a CSDF [182] graph. The choice of a MoC is up to the designer, but many signal-processing applications can be represented by dataflow graphs and the expressivity of CSDF is sufficient for the filter application. It also fits with the models of the interconnect presented in Chapter 6.

The filter, shown in Fig. 7.7a, comprises three tasks. First, the source ADC, periodically producing signed 16-bit pulse-code-modulated stereo samples. Second, the actual filter task, executed on a statically scheduled VLIW without caches. Third, the DAC, which acts as a periodic sink. Both ADC and DAC have a sampling frequency of 48 kHz, and the interconnect, processor and memory run at 48 MHz on an FPGA instance of the system. The filter task receives input samples from the ADC via the network and adds a two-tap reverberation and echo effect by mixing in past samples. The output is then sent both to the DAC and stored in the background memory for future mixing with the input. The filter application is firm real-time, as failing to consume and produce samples in 48 kHz leads to noticeable clicks in the output.

The dataflow model of the filter application is shown in Fig. 7.7b, with time indicated in network cycles (at 48 MHz). The ADC and DAC are modelled by single actors that have a response time of 1,000. The two connections, and hence four channel models, are indicated by the grey boxes in Fig. 7.7b. Every grey box corresponds to any one of the channel models in Fig. 6.5. The filter task, the processor it is running on, and its associated protocol shell, are all modelled by v_{filter}. Finally, the SRAM, the initiator bus and the protocol shell of the SRAM are modelled by v_{mem}, using the technique proposed in [187]. A detailed discussion of the last two actors follow.

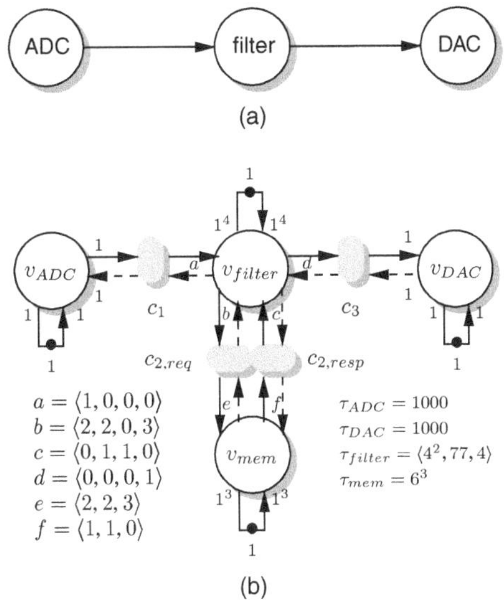

Fig. 7.7 Task graph and dataflow model of the audio post-processing filter. **a** Task graph. **b** Dataflow model

The filter actor v_{filter} has four phases, corresponding to the two read operations, the actual calculation, and the writing of output. The rates for the network channels are indicated by a, b, c and d in Fig. 7.7b. Note that the rate is the same for both the consumption (incoming edge) and production (outgoing edge). In the first phase, an input sample is read from the ADC, and a read request is sent to the memory ($h_{\text{req},r}$). In the second phase, the read response from the first phase returns one reference sample from memory ($1 + h_{\text{resp},r}$), and a new read request is sent to the memory. In the third phase, the read response from the second read returns. In the fourth and last phase, the output sample is written to memory ($1 + h_{\text{req},w}$) and sent to the DAC. The response times of the different phases are determined by manual analysis of the program flow [37]. The first two phases take four cycles each, partly due to the shell. Thereafter, 77 processor cycles are spent performing arithmetic operations and accessing local memory (without any contention). Finally, another four cycles are spent writing the output.

The memory actor v_{mem} models the shell, target bus and SRAM (no atomiser is needed as the processor always issues transactions of a single word) with a fixed access time of six cycles. The shell spends two cycles reassembling the bus transaction, and four cycles are due to pipelining of the controller and the TDM arbiter (more elaborate arbitration schemes are possible [198]). The production and consumption rates of the request and response channel are indicated by e and f in Fig. 7.7b. Reads and writes from the filter to the memory share the same connection. This is a complication, as the production and consumption rates of v_{mem} are different for the two types of operations (and also depend on the burst size). Assuming worst-case message sizes may lead to deadlock [199]. Therefore, we model the memory with three phases, corresponding to the three accesses of the filter (read, read and write). This is not a generally applicable technique, and is only possible as the order of the operations is fixed. There are, however, more elaborate dataflow models that enable variable-rate [199], but this is outside the scope of this work. In the first and second phase, a read request is received ($h_{\text{req},r}$), and a reference sample is sent back to the filter ($1 + h_{\text{resp},r}$). In the third phase, a write request is received ($1 + h_{\text{req},w}$), without sending back a response (as $h_{\text{resp}_w} = 0$). The response time for all the phases is 6 cycles, as previously discussed.

After performing the manual analysis, the filter model in Fig. 7.7b, together with a channel model from Chapter 6 is used to determine an interconnect allocation and buffer sizes such that the ADC and DAC are guaranteed to produce and consume samples in 48 kHz. Using the dataflow graph, the algorithm presented in [197] constructs a *conservative periodic schedule*. The existence of such a schedule confirms that the sink and source tasks of the filter application can indeed execute strictly periodically. Indeed, observations of the FPGA implementation confirm that the ADC and DAC do not suffer from overflow or underflow.

This simple example application, being just a pipeline with three tasks, highlights a number of important points. First, that even a very simple application like the filter requires a large amount of tedious analysis in order to derive a dataflow model. It would be virtually impossible to attempt a similar analysis for the image decoder for example. Second, how the channel model proposed in this work is easily included in a dataflow model of the application to capture the temporal behaviour of inter-task communication. Third, how the mapping to an architecture gives rise to cyclic dependencies even though the application is a pipeline. The coupling between requests and responses is inherent in memory-mapped communication, and must be captured in the model. Fourth, how cyclic dependencies are taken into account in the dataflow model. Fifth, that more elaborate dataflow models, with variable-rate, are beneficial even for very simple applications.

7.5 Conclusions

In this chapter we show how the interconnect and design flow proposed in this work is used to generate a complete multi-processor system instance, and map it to an FPGA. While this chapter demonstrates the applicability and maturity of the proposed interconnect, the target platform is not a highly-integrated low-cost SoC, but rather a system that is distributed across multiple chips and packages, costing several thousand US dollars. Moreover, since we are using an FPGA rather than an ASIC, the clock speed and area of the interconnect is roughly an order of magnitude worse than what is reported in Chapter 3. Nevertheless, despite its relatively high cost and poor performance, this case study enables us to put the problem statement and the requirements from Chapter 1 to the test.

With the FPGA case study, we demonstrate how *diverse* applications, with soft and firm real-time requirements are independently analysed, looking at different metrics, and using different methodologies. In the analysis we do not consider any of the other applications in the system. This is made possible by the *composability* of our interconnect, and is a major qualitative difference with existing MPSoC platforms. With the filter application as our example, we demonstrate how the *predictability* of our interconnect enable us to derive bounds on the end-to-end temporal behaviour. Similarly, for the decoder, we derive conservative bounds on performance for specific input images. *Reconfigurability* is used by the host to open and close connections when the different applications are started and stopped at run time, e.g. to dynamically stop the filter and replace it with the ring-tone player without affecting the other applications. Finally, thanks to the *automation* the entire system is generated in a matter of hours, based on high-level specifications and requirements.

Chapter 8
ASIC Case Study

In this chapter we exercise the proposed interconnect and design flow on two large-scale examples that have close connection to real industrial products. In contrast to Chapter 7, where we focus on the qualitative concepts and study a few applications in detail, we now turn to larger scale examples that represent state-of-the-art SoCs for consumer multimedia applications. This allows us to evaluate not only the physical and architectural, but also the *functional scalability*, i.e. the ability to scale the interconnect with increasing requirements. Using the two examples we evaluate the quantitative aspects of the proposed interconnect and also look in greater detail on the scalability of the entire design flow.

Our two examples are both high-end chips with a high degree of IP integration. The first example resembles a digital TV set-top box ASIC in the line of designs described in [61, 97]. The second example is an automotive info-tainment ASIC. In both cases, the architecture consists of one or more microcontrollers, supported by several domain- or function-specific hardware cores. As discussed in Chapter 1, this is to achieve a high computational performance at an acceptable power consumption. However, as we shall see, the communication requirements differ between the two examples, with the TV SoC having mostly high-throughput latency-tolerant connections with large burst sizes, whereas the automotive SoC has a large number of latency-critical connections with small burst sizes. As we shall see, these differences have major repercussions on the area and power consumption of the resulting interconnects.

We start with an in-depth look at the TV SoC (Section 8.1), and continue with the automotive system (Section 8.2). We end this chapter with conclusions (Section 8.3).

8.1 Digital TV

We now put the proposed design flow to use with an example inspired by NXP's PNX85500, which is a complete one-chip digital TV, aimed at the cost-sensitive midrange market. The PNX85500 delivers multi-standard audio decoding and multi-standard analogue and digital video decoding. The front-end video processing

A. Hansson, K. Goossens, *On-Chip Interconnect with Aelite*, Embedded Systems,
DOI 10.1007/978-1-4419-6865-4_8, © Springer Science+Business Media, LLC 2011

functions, such as DVB-T/DVB-C channel decoding, MPEG-2/MPEG-4/H.264 decoding, analog video decoding and HDMI reception, are combined with advanced backend video picture improvements. One of the key differentiators is the video enhancement algorithms, e.g. Motion Accurate Picture Processing (MAPP2), Halo reduced Frame Rate Conversion and LCD Motion Blur Reduction. The MAPP2 technology provides state-of-the-art motion artefact reduction with movie judder cancellation, motion sharpness and vivid colour management. High flat panel screen resolutions and refresh rates are supported, including $1{,}366 \times 768$ and $1{,}920 \times 1{,}080$ at 100 Hz / 120 Hz. The high resolution is coupled with multiplexing of multiple high-definition streams, multiple video outputs and multiple audio outputs. The SoC supports a rich set of I/O standards (e.g. PCI, USB, Ethernet, UART) and opens many possibilities for new TV experiences with IPTV and Video On Demand (VOD).

We consider the processor cores and IP blocks given, and a schematic architecture is shown in [97]. The SoC contains multiple programmable processors with caches: a MIPS for control, three TriMedia processors for audio/video, one MIPS DSP, and a 8,051 processor for power management and control. The TriMedia VLIWs are used for computation (and communication) intensive image processing, e.g. Halo Reduced Frame Rate Conversion, with on-the-fly compression of video traffic to and from the off-chip SDRAM. In addition to the processors, the SoC contains a large number of function-specific hardware IP cores. Those accelerator IPs are used for, e.g., audio and video decoding, video scaling and enhancement, graphics, and I/O (LVDS, HDMI, Ethernet, USB, Flash, SPI, PCI and I2C). Finally, the SoC has a range of peripherals for clocking, debug, timers, etc. The SoC is built around a physically centralised memory architecture. Two central off-chip memory controllers connect with two large external SDRAMs, with a 16-bit and 32-bit wide data path, respectively. The SoC contains 11 chiplets and five major frequency domains, from 27 MHz (many IPs) to 200 MHz (majority of IPs) to over 450 MHz (processors). The complete XML description of the IP architecture (corresponding to the top left architecture specification in Fig. 1.8) occupies roughly 250 lines.

A major design challenge is the on-chip communication, with 110 memory-mapped ports distributed across the IPs (the majority read-only or write-only), contributing to almost 100 logical connections. Predictability is a design constraint, and the communication requirements are given. A majority of the connections have relaxed latency requirements ($>1{,}000$ ns), as the function-specific hardware IP cores make use of prefetching read accesses and posted write accesses. The deep pipelining makes these connections latency tolerant. Processor instruction fetches (from cache misses), on the other hand, have very low latency requirements (<100 ns) and have relatively small transaction sizes, but require a limited throughput. Some accelerators, e.g. the display controller, are also latency constrained. With the most critical low-latency connections implemented through direct connections to the memory controller, and several low-throughput peripheral connections merged, this translates to 53 physical ports and 45 logical interconnections for the on-chip interconnect, all formulated in a 260 line specification (corresponding to the top right communication specification in Fig. 1.8).

Figure 8.1 shows the distribution of throughput requirements for the 45 connections, with a total of 4,402 MB/s (1461 Mbyte/s write, 2,941 Mbyte/s read). Note that this is the data throughput for the memory-mapped communication, not accounting for commands and addresses, byte masks, etc. Burst sizes vary from 64 bytes through 128 and 256 bytes, with the majority of the IPs using the latter (with a mean of 240 bytes). Including the additional address and command signals results in a raw throughput of 4,560 Mbyte/s to be delivered by the network.

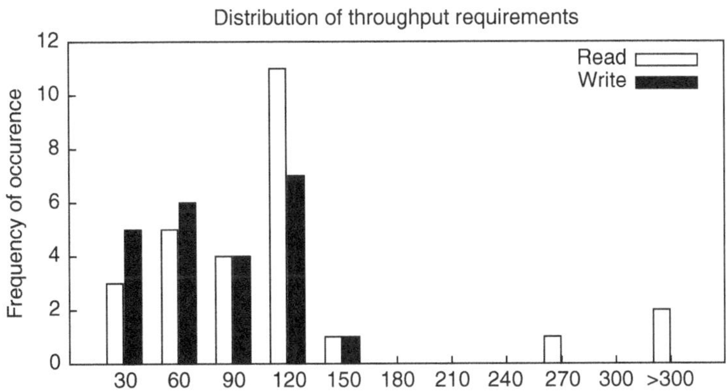

Fig. 8.1 Write and read throughput distribution (number of connections vs. throughput in Mbyte/s)

It is worth noting that this particular case study does not make use of the reconfigurability offered by the interconnect. The reason is that the specification is determined without assuming any such functionality (which is unique for NoCs). All connections are thus assumed to be concurrent, and a re-design with reconfigurability in mind could benefit from mutually exclusive applications and connections.

8.1.1 Experimental Results

As discussed in Chapter 3, topology selection is not an automated part of the design flow. The tools do, however, aid in translating a high-level topology description into an XML specification. For this example we chose a 2 by 3 mesh network (six routers) with two NIs connected to all routers but one (11 NIs in total). The router arity is thus kept at five for a single router and four for the remaining ones. This topology offers one NI for each chiplet, aiming to simplify the layout. The network is clocked at 533 MHz, which is the highest IP clock frequency.

The latency requirement, as discussed in Chapter 4, for low-latency connections is specified to 50 ns (not including the memory scheduler and controller). Given the requirements of the connections and the physical topology, the proposed design flow finds a resource allocation for the single use-case with a network slot table size of 24. The resulting allocation XML file occupies just over 1,000 lines and the corresponding architecture description is almost 4,000 lines, all automatically generated by the design flow. As the final part of the dimensioning, the NI buffers are sized using the approach from Chapter 6, and the distribution of the NI queue sizes is shown in Fig. 8.2. The large burst sizes lead to increased buffering in the NIs, and a total of 2,249 words (of 37 bits) in output queues and 2,232 words in input queues (3,060 on the initiator side and 1,621 on the target side). In total the NI buffers store roughly 24 kbyte, and the buffers have a big influence on the area requirements which we evaluate next.

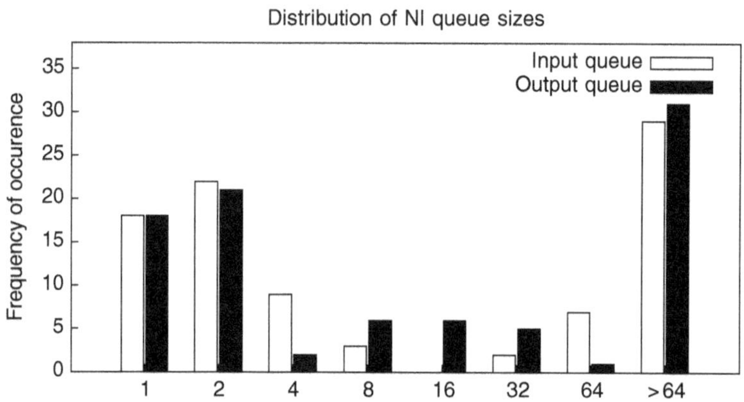

Fig. 8.2 NI queue size distribution (words of 37 bits)

Using the same synthesis setup as described in Chapter 3, i.e. Cadence Ambit with Philips 90-nm Low-Power technology libraries, the synthesised interconnect occupies a cell area of 5.5 mm^2 with 4.7 mm^2 (85%) in NI buffers. The large fraction in buffers is due to the register-based FIFO implementation, and with FIFOs based on dual-ported SRAMs the area is reduced to roughly 2.6 mm^2 with 1.7 mm^2 (65%) in buffers. Further reduction is possible with dedicated FIFOs [196], which results in a total area of 1.6 mm^2 with 0.7 mm^2 (44%) in buffers. A break down of the remaining area is shown in Table 8.1, including a comparison with area estimation models based on the synthesis of the individual blocks in Chapter 3. As seen in the table, the high-level area estimation is within 10% of the result reported by the synthesis tool, making it a useful tool for early cost evaluation. Note that the area reported is cell area only, thus before place and route, and clock-tree insertion.

Table 8.1 Area requirements

Module	Instances	Area (mm^2)	Discrepancy (%)
Router	6	0.08	+10
NI	11	0.41	−10
Shell	54	0.23	−8
Bus	6	0.08	+9

To get further insight into the area requirements and the power consumption of the interconnect instance, we synthesise the design using Synopsys Design Compiler Ultra in topographical mode. The synthesis uses TSMC's low-power 65-nm libraries (9 track with a raw gate density of 0.85 Mgate/mm^2) employing mixed Voltage Threshold (VT) synthesis with both high- and low-VT libraries. The synthesis is performed on a flattened design with multi-stage clock gating enabled, using integrated clock gates from the technology library. The power minimisation is driven by activity traces captured in RTL simulation using traffic generators to mimic the IP behaviour.

The total area after synthesis is estimated at 2.5 mm^2 (9,781 cells), which is to be compared with the 5.5 mm^2 using the 90-nm libraries. Once again the FIFO implementation used is the register-based FIFO, leading to a dominant part of the area spent in buffering. The area of the buffers is, however, reduced by the clock gating as many multiplexers in the FIFOs can be replaced. A total of 7,500 clock gating elements are inserted, resulting in 97% of the 25,0000 registers in the design being gated. At a global (worst-case) operating voltage of 1.08 V the cell power is estimated at 80 mW and the net switching power at 43 mW. The total dynamic power, under worst-case conditions, is 123 mW and the leakage power is estimated at 3 mW (or roughly 2.5% of the total power consumption). Roughly 22% of the dynamic power is spent in the clock tree. We can conclude that the low-power libraries, in combination with mixed-VT synthesis leads to an implementation where leakage-minimisation techniques like power gating, which is costly in level shifters, *are not worthwhile*.

Next, we evaluate the behaviour of the generated interconnect hardware and software by simulation, using an instance of the entire system with traffic generators to mimic the IP behaviour. Instrumentation is (automatically) inserted in SystemC and VHDL testbenches respectively to observe the utilisation of the different modules, the latency of individual packets in the network, and the end-to-end latency and throughput between the IPs. The cycle-accurate SystemC simulation executes with a speed of roughly 50 kcycles/s on a modern Linux machine (1 ms of simulated time in less than 20 s). The corresponding VHDL testbench is roughly 200 times slower, with 0.2 kcycles/s. Simulation of the synthesised netlist is again a factor 100 slower, reducing the speed to roughly a cycle per second. Note that the throughput and latency of all connections is guaranteed, but the simulation results give additional confidence and insight in the actual average throughput and latency for the particular use case.

First, looking at the critical low-latency connections (specified to 50 ns as previously mentioned) they consistently have an average network latency of 19 ns and a maximum across all the connections of 42 ns. For a specific low-latency connection, the end-to-end latency for a 128 byte read transaction, including the memory controller and memory access, is 280 ns on average and 308 ns for the observed worst case (corresponding to 190 processor cycles at 533 MHz). All the 45 connections have an observed throughput and latency that meets the user-specified requirements, and this behaviour is consistent in both SystemC and VHDL simulation. Comparing SystemC, behavioural VHDL and netlist simulation we see slight differences in the statistical traffic profile. The reason for the discrepancies is a the use of different traffic generators for the SystemC and VHDL testbenches. The results, however, looking at the end-to-end latency and throughput, are within ±3%. The simulation results confirm that the requested performance is delivered.

Next, we look at the utilisation of the interconnect, as observed during simulation. The TDM-based arbitration of the network requires low-latency connections to reserve a larger number of appropriately spaced time slots. As a consequence, these slots are often unused. The utilisation does hence not reflect the reserved slots, but rather the slots that are actually used at run time. We first look at the ingress and egress links of the network, i.e links that connect an NI to a router. Since the NI is comparatively more expensive, the chosen topology aims to minimise the number of NIs. The ingress and egress links are therefore more likely to have a high utilisation compared to the average over the entire network. For the TV SoC, the average NI link utilisation is 49%, with a minimum of 27% and a maximum of 64%. This is to be compared with an overall network link utilisation of 22%. The link with the lowest utilisation is only used 12% of the time. The relatively low utilisation of the network stresses the importance of using clock gating as part of the logic synthesis. Register-level and module-level clock gating ensures that the power consumption is proportional to the usage of the gates rather than the sheer number of gates. As already mentioned, the total interconnect power consumption, under worst-case conditions and with register-based FIFOs, is estimated at 125 mW.

8.1.2 Scalability Analysis

Having evaluated our proposed interconnect in the context of the contemporary digital TV SoC, we now continue to look at how the interconnect scales with the requirements, and how suitable it is for future generation multimedia applications. As discussed in [97], the number of latency-critical connections has stayed low and fairly constant during the past six generations of TV SoCs in the Nexperia platform. The total throughput, however, and the number of latency-tolerant function-specific hardware IPs grow over time. These key trends are a result of stable functional subsystems being moved to latency-tolerant implementations as the functionality matures. The latter is a necessary trend to fit more functionality on a single SoC and still be able to serve the latency-critical connections within acceptable bounds. The on- and off-chip throughput is scaling with the generations, but the latency (in absolute time) is fairly constant, leading to growing memory access latencies (measured in processor cycles). With a growing number of latency-critical connections the communication pattern would be inherently non-scalable.

Our first scaling experiment fixes the IPs and their 45 connections, and only scales the throughput requirements (leaving the latency untouched). For each design point, we re-run the entire interconnect design flow using the new (scaled) requirements. For each point, we evaluate two different network topologies. First, a minimum mesh, determined by a simple loop that increases the dimensions until a feasible resource allocation (one that satisfies all requirements) is found. For simplicity, the number of NIs is fixed at three per router. Second, a point-to-point topology where each connection is mapped to two unique NIs interconnected by a router (to allow for instantiation of the resource allocation during run-time). A maximum of 32 TDM slots is allowed in both cases. Once a resource allocation is found, we determine sufficiently large buffer sizes (also verifying the performance bounds using dataflow analysis) and estimate the required area using the models from Chapter 3. Thus, for every point we determine two complete hardware and software interconnect instances that satisfy all the requirements.

Figure 8.3 shows the area, both with and without buffers, for the two topologies as the throughput requirements are scaled from 0.2 to 3.4 times the original TV SoC. The area estimates once again assume a 90-nm process and register-based FIFOs and are thus overly conservative. However, as we are interested in the scaling the absolute numbers are not critical. What is more interesting is the trend as the scaling factor is increased. For scaling factors below three, the total area is growing linearly for both topologies, with the area of the mesh being roughly 25% smaller than the point-to-point topology (75% smaller when excluding the buffers). Although not visible in the figure, the mesh area excluding the buffers is increasing in a stair case as the number of routers and NIs grow. Increased throughput leads to different mapping and routing decisions as the connections are balanced on a larger interconnect. As expected, the point-to-point topology area is constant when the buffers are not considered. With scaling factors above three, the mesh area is increasing faster, approaching the area of the point-to-point topology, and after 3.4 no feasible resource allocation is found for either of the two topologies. This is due to the inherent sharing of the IP ports where the scaled requirement eventually saturate the capacity. Thus, the point of contention is pushed to the IPs ports and there exist no possible topologies and resource allocations, for any NoC architecture or design flow.

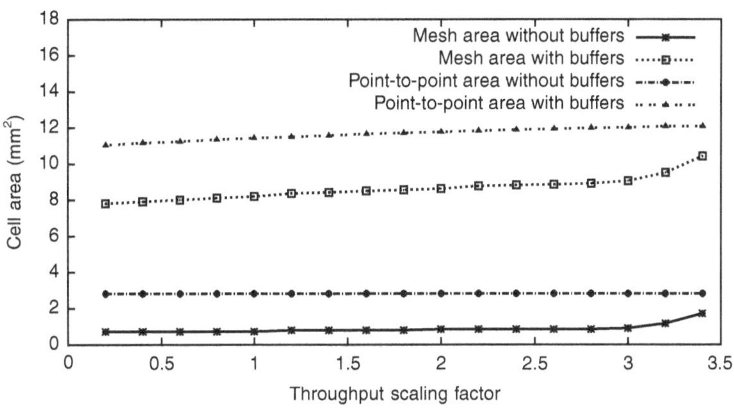

Fig. 8.3 Scalability of throughput requirements

We have seen that the interconnect scales linearly with the throughput requirements when the number of connections is kept constant. Now we look at how the interconnect scales with the number of latency-tolerant connections. Rather than extrapolating a next-generation TV SoC, we randomly select a subset of the connections of the original example, and do so for fractions ranging from 0 to 1 in increments of 0.1. For each design point (belonging to a specific fraction) we repeat the random selection 10 times. As a result, for each point we re-run the entire design flow 10 times with (potentially) different subsets of the communication specification. For each run we determine the minimum mesh as before. After finding a feasible resource allocation the buffers are sized and the interconnect area estimated.

Figure 8.4 shows the area estimates with and without buffers, with each point showing the average of the 10 runs together with the standard deviation. Once again, the absolute numbers are not particularly important as they depend heavily on the chosen technology node and buffer implementation. What is important is that both Fig. 8.4a, b, show a clear linear trend. This suggests that the interconnect architecture and design flow is providing functional scalability and thus

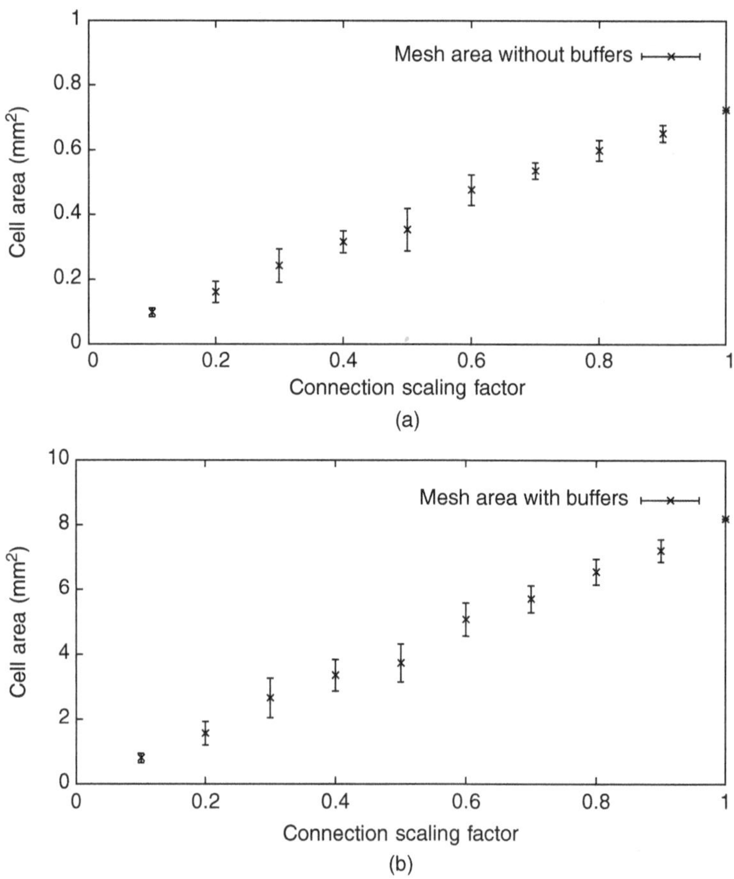

Fig. 8.4 Scaling the number of connections. (**a**) Area without buffers. (**b**) Area with buffers

able to accommodate a growing number of connections with a roughly constant cost/performance ratio.

8.2 Automotive Radio

Our second example is a software-defined radio SoC, aimed at the automotive industry. Although mobility, connectivity, security and safety drive the automotive industry, there is an accelerated adoption of consumer features in cars, with home entertainment also in your car. The main applications include high-quality, real-time, interactive audio and video, with standards like DAB, DVB-H, DVB-T, T-DMB, HD Radio, etc. While delivering a true multimedia experience, the SoC must also meet automotive quality demands. Due to the nature of the SoC, there is an essential need for a powerful and flexible multi-standard, multi-channel software-defined radio, with efficient channel decoder processing.

Similar to the digital TV SoC, we consider the processor cores and IP blocks given. The example is inspired by a SoC contains three programmable processors with caches: an ARM Cortex for control, and two Embedded Vector Processors [22] (EVP). The EVPs constitute the core of the software-defined radio. In addition to the processors, the SoC contains a large accelerator subsystem. The accelerators are involved in signal-processing and I/O (several DACs and ADCs, DMA and USB). In contrast to the digital TV SoC, this system is not making use of a physically centralised memory architecture, as it has many on-chip (more than 10) SRAMs in addition to the external SDRAM. As we shall see, the result is less inherent contention in the communication patterns. The SoC contains 7 chiplets and 9 clock domains, ranging from 100 to 350 MHz with the majority of the IPs running at 100 or 200 MHz.

Communication is central to the SoC with 68 IP ports using both memory-mapped and streaming communications with a total of 52 logical connections. Many of the IPs do not use posted/pipelined transactions and their communication is therefore latency-critical. In contrast to the digital TV it is thus not only cache traffic that is latency critical. Due to the real-time nature of the application, predictability is crucial for all on-chip communication. After merging a few low-throughput connections to peripherals, the on-chip interconnect must accommodate 43 ports (distributed across 12 IPs), with 32 logical interconnections. Roughly half of these connections have latency requirements below 100 ns.

Figure 8.5 shows the distribution of throughput requirements, with a total of 2,151 Mbyte/s (997 Mbyte/s write, 1,154 Mbyte/s read). Including command and address this amounts to 4,183 Mbyte/s to be delivered by the network. Note that this is almost the double due to the small burst sizes, varying between 4 and 128 bytes (mean of 31 bytes). Compare this to the digital TV where we almost exclusively have large bursts, making the raw throughput requirement of the two case studies very similar. Although the throughput requirements are not much different, we shall see that the burst size has a tremendous impact of the NI buffers, and consequently the interconnect area and power.

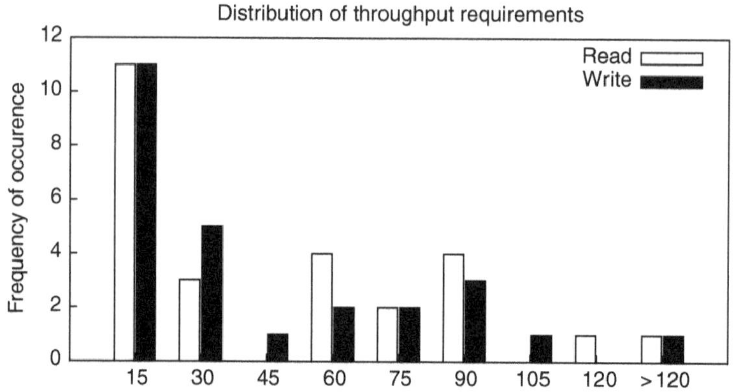

Fig. 8.5 Write and read throughput distribution

Similar to the TV SoC, this case study does not exploit the support for multiple use-cases.

8.2.1 Experimental Results

For the automotive SoC, we use a 3 by 3 mesh (nine routers) with two NIs connected to the corner routers, one NI for the remaining peripheral routers, and no NI for the central router (12 NIs in total). The router arity is thereby kept low and the topology is still large enough to enable the IPs to be placed around the interconnect, grouped according to chiplets and clock domains. For this case study, the network is clocked at 400 MHz which is double the frequency of most IPs.

Given the requirements, an interconnect instance is generated with a slot table size of 21 and the buffers are sized according to the proposed design flow. The distribution of NI queue sizes is shown in Fig. 8.6, with a total of 442 words in input queues, and 587 words in output queues (453 on the target side and 576 on the initiator side) the total capacity is roughly 5 kbyte. Compared to the TV SoC this is roughly four times less buffering per connection, largely due to the big difference in average burst size.

The difference in buffer size also has a big impact on the interconnect area. Using the same 90-nm libraries and the same setup as for the TV SoC, the total cell area for the automotive SoC interconnect is 2.13 mm^2 with 1.41 mm^2 (66%) in register-based NI buffers. With dual-ported SRAMS this is reduced to 1.22 mm^2 with 0.5 mm^2 (40%) in buffers, or as little as 0.9 mm^2 with 0.22 mm^2 (23%) in buffers when implemented using dedicated FIFOs [196]. An area break down is shown in Table 8.2, including both the results from synthesis and the high-level area-estimation models. Just as for the TV SoC, the area estimation is within 10%, suggesting the models can be used reliably for early design-space exploration and evaluation.

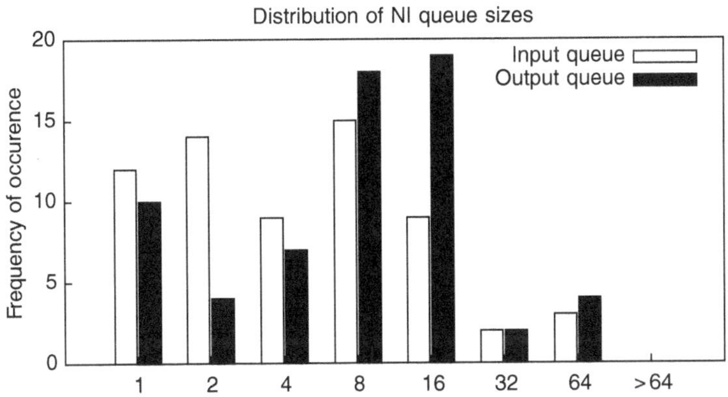

Fig. 8.6 NI queue size distribution (words of 37 bits)

Table 8.2 Area requirements

Module	Instances	Area (mm^2)	Discrepancy (%)
Router	9	0.13	+10
NI	12	0.32	−8
Shell	43	0.18	0
Bus	8	0.07	−4

We synthesise also this interconnect using Synopsys Design Compiler Ultra, aiming to minimise the power consumption. The details are the same as described in Section 8.1. The total area after synthesis is estimated at 0.78 mm^2 (4,784 cells) and 90% of the 76,474 registers are clock gated. At worst-case conditions, the cell power is estimated at 41.2 mW and the net switching power at 15.4 mW, resulting in a total dynamic power consumption of only 57 mW. The leakage is reported to be less than 1 mW, once again demonstrating the abilities of the low-power libraries and mixed-VT synthesis. Roughly 27% of the total power is spent in the clock tree.

Similar to the TV SoC, we perform SystemC and HDL simulations on the instantiated interconnect. Once again, all throughput and latency requirements are satisfied, and this behaviour is consistent for all the different instantiations. Looking at the utilisation during simulation, the automotive SoC has an average NI link utilisation of 30% with a minimum of 7% and a maximum of 75%. The low utilisation of the NI links is due to the very strict latency requirements which limits the amount of sharing. The overall network link utilisation is also low, only 20%. As previously discussed, the relatively low utilisation highlights the importance of clock gating, as used in our experiments.

8.2.2 Scalability Analysis

We continue our evaluation with two different scaling experiments, similar to what is described for the TV SoC in Section 8.1. First, scaling the throughput requirements between 0.2 and 4.8, we compare the area of a minimum mesh and a point-to-point topology. Figure 8.7 shows the result, both with and without buffers. The heuristics used in the mesh sizing results in an area that is growing in a stair case, which is now visible as the total area is less dependent on the buffers. The total area is growing linearly for both topologies until they saturate at a scaling factor of 4.8 where the requirements can no longer be allocated. At this point the mesh area is almost equivalent to the point-to-point topology, using a 5 by 4 mesh with 3 NIs per router. Comparing the results with the TV SoC we clearly see the consequences of having a more distributed communication pattern (more memories) as the throughput can be scaled further (4.8 compared to 3.4).

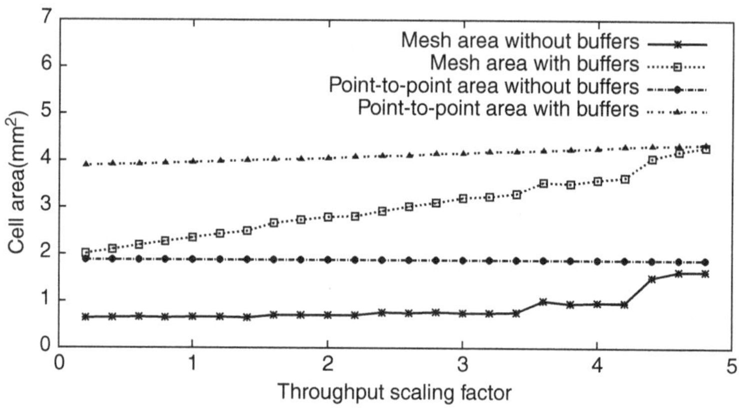

Fig. 8.7 Scalability of throughput requirements

Next we evaluate the scalability with respect to the number of connections. In contrast to the TV SoC we now include all connections (latency critical and latency tolerant) in the selection. Again we look at 10 different design points and run the entire design flow 10 times per point. Figure 8.8 shows the area with and without buffers. Although there is a considerable variation in the results, the linear trend is clear. We have thus shown that also for this example, with a larger number of latency-critical connections, but a more distributed communication pattern, the design flow offers functional scalability.

8.3 Conclusions

In this chapter we demonstrate the use of the proposed interconnect and design flow using two large-scale industrial case studies. In contrast to previous chapter, we evaluate the quantitative aspects of the interconnect with focus on the *scalability*. Rather than targeting an FPGA, we look at the performance and cost in the context of an ASIC flow, targeting modern CMOS technologies. For the two case studies, we show the large impact burst sizes have on the interconnect area and

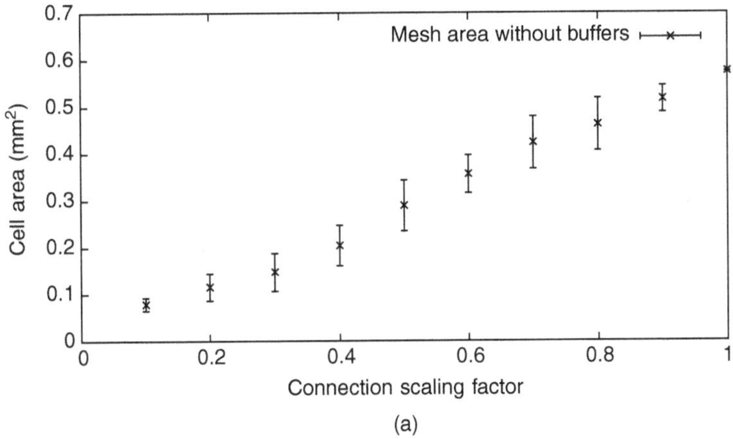

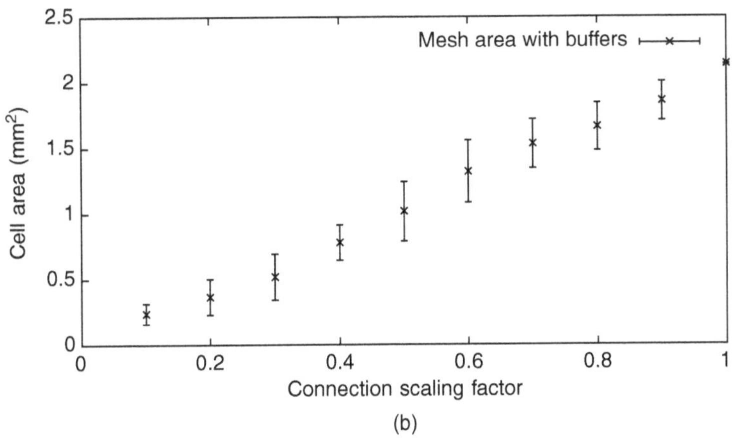

Fig. 8.8 Scaling the number of connections. (**a**) Area without buffers. (**b**) Area with buffers

power consumption, and quantify the influence latency-critical connections have on the network utilisation. Even with a low utilisation, our interconnect achieves a competitive performance/area. The utilisation is, however, central for the performance/power trade off, where it determines what kind of low power technique to apply. The relatively low utilisation (around 25%) stresses the importance of clock gating in the logic synthesis to minimise dynamic power. We show that low-power mixed-voltage-threshold libraries lead to leakage power being less than 3% in a 65-nm technology. To evaluate the functional scalability, we generate many hundreds of application-specific interconnect instances based on our two case studies, and show that the proposed interconnect is able to accommodate a growing number of requirements with a constant cost/performance ratio.

Chapter 9
Related Work

After completing our detailed exposition of the proposed on-chip interconnect, we now look at related work, and how they address the requirements from Chapter 1. Throughout this chapter, we highlight the contributions of our proposed interconnect and how it compares to the related works. For a more detailed exposition of the key concepts, we refer back to Chapter 2.

9.1 Scalability

We divide the discussion on scalability in two parts. First, we look at how different interconnects address physical scalability, i.e. the ability to scale the interconnect to larger die sizes without impairing the performance. Thereafter we raise the level and look at architectural scalability, i.e. the ability to add more applications and IPs without negatively affecting the performance.

9.1.1 Physical Scalability

Traditional *single-hop* interconnects have a limited physical scalability as they introduce long global wires [43]. *Multi-hop* interconnects, spanning from segmented buses [36, 159] to NoCs [19, 43, 173], address the wire-length issue by splitting the interconnect into multiple (pipelined) segments. However, even with a NoC, the clock distribution scheme limits the physical scalability.

Many networks [63, 108, 122, 157] rely on a (logically) *globally synchronous* clock. While techniques such as link pipelining have been proposed to overcome link latency [162, 184], the cycle-level synchronicity severely constrains the clock distribution [200] and negatively affects the scalability [29].

On the other end of the spectrum, a range of *asynchronous* networks [12, 27, 165] completely remove the need for a clock, and aim to leverage the characteristics of asynchronous circuits to reduce power and electromagnetic emissions. While offering physical scalability, these networks rely on not yet well-established design

A. Hansson, K. Goossens, *On-Chip Interconnect with Aelite*, Embedded Systems,
DOI 10.1007/978-1-4419-6865-4_9,

methodologies and verification techniques. They also complicate the provision of composable and predictable services.[1]

To overcome the disadvantages of global synchronicity, but still enable a traditional synchronous design style, the networks in [29, 113, 154, 200] use *mesochronous* and asynchronous links [124] between synchronous network elements. By allowing local phase differences, clock distribution and placement is greatly simplified [200]. Moreover, the global phase difference is allowed to grow with chip size. In fact, mesochronous network interconnecting many tens of processors have already been demonstrated [67, 83, 194]. An additional benefit of mesochronous network is that they can be perceived as synchronous on the outside of the network [29]. This greatly simplifies the provision of composable services.

The constraints on the clock distribution within the interconnect apply also to the (global) control infrastructure. Thus, dedicated interconnects [115, 203] that rely on global synchronicity impair the overall physical scalability and must be avoided. By reusing one and the same interconnect for both user data and control, as proposed in [46, 68, 157], the problem is reduced to providing one scalable interconnect.

In addition to relaxing the constraints on the clock distribution within the interconnect, it is crucial for physical scalability that IPs are able to use independent clocks (from each other as well as the interconnect). While asynchronous networks natively offer a GALS design approach at the IP level, e.g. through the use of pausable clocks [13], synchronous and mesochronous networks must add clock domain crossings at the interfaces to the IPs. Solutions based on bi-synchronous FIFOs, placed inside the NIs, are proposed in [28, 63, 154]. Such a clock domain crossing, e.g. [196] that is used in [63], is robust with regards to metastability and offers a simple protocol between synchronous modules, while still allowing every module's frequency and voltage to be set independently [103].

In this work we extend on the logically synchronous Æthereal [63] network by offering mesochronous links [78]. Similar to [46, 68] the same interconnect is used for both data and control, but we introduce dedicated control buses that allow the overall control infrastructure to be used in a system where the IPs run at different clock frequencies. Moreover, with respect to [168], we move the clock domain crossings outside the NI, thus improving modularity and greatly simplifying the design.

9.1.2 Architectural Scalability

Moving from the physical to the architectural level, scalability is the ability to accommodate a growing number of applications and IPs, without compromising their performance. Similar to the physical scalability, single-hop interconnects limit also the architectural scalability. In bus-based interconnects, e.g. the Silicon Backplane [202], the performance decreases as the amount of sharing increases. In

---

[1] A globally asynchronous implementation of our proposed router network is possible [52, 78] without any change to the concepts, but the uncertainty pertains the length of a slot (in absolute time), determined by all the routers together.

crossbar-based interconnects, e.g. Prophid [107], a growing number of IPs lead to a rapid growth in silicon area and wire density. This in turn causes layout issues (e.g. timing and packing) and negatively affects performance [162]. Although the problem can be mitigated by using a partial crossbar [135], the key to architectural scalability is a distributed multi-hop interconnect.

In a multi-hop interconnect, e.g. Sonics MX [179] that uses a hybrid shared-bus and crossbar approach, the individual buses and crossbars are not scalable on their own. Rather, the architectural scalability comes from the ability to arbitrarily add more buses and crossbars without negatively affecting the performance of other parts of the interconnect. A segmented bus [36, 159] achieves architectural scalability by partitioning a bus into two or more segments, operating in parallel, and connected by bridges. NoCs extend on the concepts of segmented buses and offer a more structured approach to multi-hop interconnects [19, 43].

From the perspective of architectural scalability, there is an important distinction to be made between NoCs where arbitration is performed at every hop (routers), e.g. [12, 27, 154, 162, 165] and those where arbitration only takes place only at the edges (NIs) of the network, as done in Nostrum [122], aSoC [108], Æthereal [63], TTNoC [157], and our proposed network. The works in [12, 27, 154, 162, 165] use virtual channels to provide connection-based services. As a consequence, the router area grows quadratically with the number of connections passing through it. Moreover, the maximum frequency of the router is also reduced, and the router is likely to be part of the critical path of the network. In contrast to the NoCs that rely on virtual channels, the NoCs in [63, 108, 122, 157] as well as our proposed interconnect all have *stateless* [186] routers. Thus, the router is independent of the number of connections in terms of area (number and depth of buffers) and speed (arbitration). This moves the router out of the critical path and pushes the issue of scalability to the NIs. As shown in Chapter 3, the scalability of the NI is negatively affected by the size of the TDM tables (which depends on the number of connections and their requirements), but can be heavily pipelined.

In this work we extend on the architectural scalability of Æthereal [63] by offering a significantly smaller and faster router design [78]. This is achieved by excluding best-effort services. Moreover, extending on [168], memory-mapped initiator ports that use distributed memory communication may use different NIs for the connections to different targets. Similarly, a memory-mapped target port that is shared by multiple initiators may distribute the connections across multiple NIs. Thus, it is possible to push the point of contention all the way to the shared target port and offer arbitrarily low latency and high throughput in the interconnect.

9.2 Diversity

Applications and IPs in a system typically come from different providers and aim to provide widely varying services to the user. Thus, diversity is important on all levels ranging from application requirements, down to the physical interfaces of the IPs.

Much work on NoCs is focused on the router network and does not address communication at the IP level. For example, networks with adaptive routing [153] typically ignore the ordering even within a connection and it is unclear how to (efficiently) use such a network to interconnect IPs.

NI architectures that provide connection-level services are presented in [28, 32, 150, 157, 168, 184, 201]. A specialised NI, tightly coupled to an accelerator, is implemented and evaluated in [32]. The NI in [157] takes a slightly more general approach, offering time-triggered exchange of application-level messages. However, the problem of flow control, even at the packet level, is pushed to the IPs. Communication must consequently take place at a priori-determined instants [99], placing many constraints on the behaviour of the IP. Streaming-like and memory-mapped communication at the level of connections is provided by [168] and [28, 150, 184, 201], respectively. The NIs in [28, 150, 156, 184] interface with the IPs using OCP interfaces. However, they only use OCP as a data-link level protocol and ignore higher-level protocol issues like ordering between connections. Only [168, 201] address distributed and shared memory communication. However, neither of the works discusses the impact of memory-consistency models on the protocol stack.

With more elaborate programming models, it is also necessary to address the issue of message-dependent deadlock [178]. Most NoCs rely on strict ordering with separate physical or logical networks, thus severely limiting the programming model (e.g. to pure request-response protocols). End-to-end flow control is proposed in [73] to avoid placing any restrictions on the dependencies between connections outside the network.

In this work, we consider event-triggered communication with explicit flow control, thus enabling a wide range of applications. The interconnect is optimised for streaming communication but offers also distributed shared memory-mapped communication, with well-defined memory-consistency models and no restrictions on the programming model. Extending on [63, 168], we clearly separate the network stack, the streaming stack and the memory-mapped stack. This is accomplished by unifying the ports on the NIs, by limiting the functionality of the protocol shells to the data-link level (and thus completely decoupling message and packet formats), and by extending the network with buses.

9.3 Composability

Composability is a well-established concept in systems used in the automotive and aerospace domains [6, 98, 169]. An architecture is said to be composable with respect to a system property if system integration does not invalidate this property once the property has been established at the subsystem level [98]. In this work, similar to [77, 90, 101, 109, 143, 169] the property we look at is the temporal behaviour. There are, however, large differences in the level of composability, the mechanisms used to enforce it, and the level of interference allowed. We now discuss these three topics in greater depth.

9.3.1 Level of Composability

In [101], composability is provided at the level of components (what we refer to as tasks). Composability of tasks, also within an application, limits the scope to applications for which a static schedule (for all tasks of all applications) can be derived at design time, and the interfaces of the tasks (components) are fully specified in the value domain and in the temporal domain [99]. Such a specification is not feasible for, e.g., for dynamic applications like a video decoder or I/O interfaces. Thus, in contrast to [101], this work provides temporal composability of applications (what is called a disjoint subgroup of cooperating components in [101]) rather than tasks.

Despite the difference in level of composability, the principles of composability, as suggested in [101], still apply. At the application level, the first principle, *interface specification*, is trivially met as we currently limit ourselves to no (temporal) interactions between applications. Note, however, that non-consumable reads and writes, i.e. non-interfering interactions, are allowed. That is, one application may (over-)write output in a buffer that is read (sampled) by another application. An example is shown in Fig. 1.1, where the MPEG-1 decoder application produces samples using, e.g., a double-buffering scheme, and the most recent sample is read by the audio post-processing task. The second principle, *stability of prior services*, is provided at the level of applications. Thus integrating a new application does not affect other applications, as long as their allocations remain unchanged. The third principle, *non-interfering interactions*, is the very foundation of our proposed interconnect as it eliminates all interference between connections and hence also applications. The fourth principle, *error containment*, is also provided at the level of applications as there is no possibility of an error in one application affecting another. Note, however, that we currently do not enforce spatial containment (i.e. fixed regions) at shared target ports. The last principle, *fault masking*, is something we consider to be orthogonal to composability (when provided at the application level).

9.3.2 Enforcement Mechanism

In the automotive and aerospace domains composability is traditionally achieved by not sharing any resources [169]. This approach, however, is much too costly in the consumer-electronics domain, where resources, such as accelerators, interconnect and memories, are shared between applications [50]. Nostrum [122], Æthereal [63] and TTNoC [157] offer TDM-based composable (and predictable) services at the level of individual connections. The TTSoC architecture extends those services to the task level by making the entire system time-triggered. The complexity of interactions between tasks (and thus applications) is moved to upholding and adhering to the coarse global time base and the global schedule. The responsibility of doing so is pushed all the way to the IPs, e.g. data is lost if the IPs do not adhere to the schedule. A similar approach is proposed in [109], where the IPs are responsible for all contention resolution.

In this work, in contrast to [109, 157], we do not base composability on the concepts of a time-triggered architecture, but instead use an *event-triggered architecture with budget enforcement* [14] that does not place any restrictions on the applications. Inside the network and at shared targets, we completely remove the uncertainty in resource supply by using *hard resource reservations* [163], where the amount of service and the time at which an application receives its service is independent of other applications. We use the concept of contention-free routing [164] to provide composable sharing of the network at the level of connections. In contrast to [157] we place no responsibilities on the network end-points. Extending on [63, 168] we enable composable sharing of memory-mapped target ports. However, in our current implementation, we do not allow sharing of processors between applications (work in progress).

9.3.3 Interference

Composability, as defined in this work, does not allow any application interference at all, down to the clock cycle level. This is different from the TTSoC architecture [101], where interference (between tasks) is allowed as long as it is within a coarse-grained time step, bounded both from above and below. Thus, jitter is allowed at a fine-grained level, as long as the characterisation is still valid on the coarser grain. The restrictions on interference are further relaxed in [90, 114, 143] that only bound interference from above and not from below.

As an implication of *removing rather than limiting interference*, the capacity unused by one application cannot be given to another one. The reason for our strict definition is that, in the general case, it is not possible to say what effects a small perturbation at the task level would have on the complete application. If *work-conserving arbitration* is used between applications [143], even temporarily [114], the application composability is lost. A task can be given *more resources*, and allocated resources *at an earlier point in time*, both *as the result of another application's behaviour*. While the additional resources or earlier service might seem positive at first, for a general application, more resources does not always result in a improved quality of service. More resources might *trigger bugs*, e.g. due to races that are schedule dependent. Additionally, *scheduling anomalies* [64] lead to situations where an earlier service for an individual task may lead to reduced performance at the application level. It is also possible that *schedule-dependent decisions* inside an application cause a reduction in quality. Even an improved deadline miss rate is not always positive, as rapidly *changing quality levels* are perceived as non-quality [1, 31, 204]. All the aforementioned effects depend on the other applications in the system. In other words, verification is *monolithic*, unless all applications are free of errors and have a known (and characterised) worst-case behaviour and fit in a monotonic model of computation.

In this work, we completely remove all application interference. Capacity unused by one application is not given to another one, thus enabling temporal application composability without placing any requirements on the applications. Unused capacity can, however, be distributed within one application at the channel trees, by distinguishing inter-application and intra-application arbitration [71].

9.4 Predictability

We split the discussion of predictability into three aspects: the mechanism used to enforce a specific temporal behaviour in shared resource, the allocation of these resources to the applications, and the analysis of the mechanism and allocation together with the applications.

9.4.1 Enforcement Mechanism

Many NoCs provide latency and throughput bounds for one or more connections [12, 26, 63, 95, 108, 122, 154, 157, 165, 193]. Most common are guarantees based on virtual circuits [12, 27, 154, 165]. With strict priority arbitration in the routers [12, 154, 165], only one virtual channel per link can be given bounds on its latency and throughput, due to the lack of rate regulation. Hence, connections cannot share links. The Mango NoC overcomes the problems of strict priority-based arbitration by introducing a rate-regulator [27]. A similar approach, albeit with different scheduling mechanisms is used in [95] and [193]. As we have already seen, Nostrum [122], aSOC [108], Æthereal [63] and TTNoC [157] implement the guarantees by globally time-multiplexing the communication channels.

In this work, we use the concept of contention-free routing [164] to enable bounds on latency and throughput in the network. In contrast to other on-chip interconnects, we offer predictability not only in the network, but also at shared target ports through the insertion of buses and atomisers.

9.4.2 Resource Allocation

With the ability to offer throughput and latency guarantees in the network, it remains a problem to allocate the resources for the different connections. The NoC architecture must be coupled with tooling that enables a designer to go from high level requirements to resource allocations. Many works only demonstrate the provision of latency and throughput guarantees for a handful connections [12, 27, 122, 154, 165] (and some works none at all). For the networks where a large number of connections have been demonstrated [63], the automation is focused on throughput and does not discuss the provision of latency guarantees [70–72].

In this work, we guarantee both the throughput and latency requirements of individual channels are satisfied. Moreover, we demonstrate the ability to do so for many hundreds of connections, with widely varying requirements.

9.4.3 Analysis Method

Even with bounds on throughput and latency *inside the network*, to enable predictable temporal behaviour at the application level, it is necessary to *include the application* and the buffers between the application and the NoC in the performance analysis [79, 82].

Simulation is a common approach to performance analysis [135]. Trace-based buffer sizing provides an optimal bound on the buffer size for the given input traces. However, it does not guarantee that the derived size is sufficient for other traces, i.e. for other traces the performance requirements might not be met and the application might even deadlock. The algorithm in [39] uses *exhaustive simulation* of given periodic traces. While the method provides tight bounds, it does so at the price of a high run time, requiring hours or even days for larger SoC designs. Moreover, the applications, as well as the NoC, are assumed to be completely periodic, thus severely limiting the scope.

More generally applicable than exhaustive simulation is to use conservative linear bounds on the production and consumption of data [56]. Assuming it is possible to find such a *traffic characterisation*, the buffers are sized so that they are never full. The coarse application model results in a low run time of the analysis, at the cost of large buffers. The restrictive application model also limits the scope of applications, and it remains a problem to derive a conservative traffic characterisation for a given application. Moreover, both the aforementioned approaches [39, 56] are unable to determine the temporal behaviour for given buffer sizes, which may be necessary if an application is mapped on an existing architecture.

In [75, 82, 126] the application, as well as the NoC, is modelled using dataflow graphs. Thereby, in contrast to [39, 56], it is possible to either compute buffer sizes given the application requirements or derive bounds on the temporal behaviour (latency and throughput) for given buffer sizes [15, 131, 182]. The latter is demonstrated in [82, 126] where applications are mapped to an existing NoC platform. However, [82, 126] do not discuss what NoC properties are necessary to derive such a model or how it can be applied to other NoCs. Additionally, neither of the works compare the dataflow analysis with existing approaches for buffer sizing.

In this work, we give a detailed exposition on how to construct a dataflow graph that conservatively models a NoC communication channel, and the relation between the architecture and the model. We demonstrate how the model can be used to determine buffer sizes for given requirements, and to derive guarantees on the temporal behaviour of an actual application.

9.5 Reconfigurability

A comprehensive model of dynamic change management is given in [102]. The work discusses the requirements of the configuration management and the implications of evolutionary change of software components. A practical but more limited approach to dynamic application reconfiguration in multi-processor SoCs is presented in [93, 170]. Reconfiguration of the interconnect is, however, not addressed.

Much work is focused on complete NoC design and compilation flows [23, 62, 111]. While the works suggest automated generation of control code [62] and standardised NoC application programming interfaces [111], no details are given on how these dynamic changes are safely implemented.

Methodologies for dynamic run-time reconfiguration of NoCs are presented in [145, 176]. Both works assume little or no design time knowledge about the applications and defer mapping decisions to run time. While offering maximal flexibility, guaranteeing that an application's requirements are met is difficult due to possible resource fragmentation over time. Mitigating the problem necessitates complex migration schemes with unpredictably large delays [145] as applications are interrupted during migration. In [72], multiple use-cases are allocated at compile time, enabling guarantees on seamless starting and stopping of a given set of applications. However, while showing how to determine allocations, the work does not show how to deploy them at run time.

The works in [46, 68, 115, 203] describe how to implement a control infrastructure for network reconfiguration. A dedicated control interconnect is used in [115, 203], whereas the functional interconnect is used to carry control data in [46, 68]. Limited or no details are given in [46, 115, 203] of how to actually use the infrastructure to ensure a consistent state after, as well as during, reconfiguration. The actual detailed reconfiguration process is investigated in [68] and a library for NoC reconfiguration is presented. None of the aforementioned works [46, 68, 115, 203] provide temporal bounds on the reconfiguration operations.

In this work, extending on [68], we show how to efficiently use the guaranteed services of the NoC to implement a virtual control infrastructure as two channel trees [71]. Moreover, we show how this enables temporal bounds on the reconfiguration operations. Through the structured addition of control buses, we provide a scalable and generic control infrastructure. In contrast to [145, 176] we assume that the mapping of tasks to IPs is fixed, and do not allow run-time reallocation of resources.

9.6 Automation

A large body of works on interconnect automation focus on the problem of mapping IPs onto the NoC topology and routing of the connections [85, 86, 133, 134, 136]. In [85] a branch-and-bound algorithm is used to map IPs onto a tile-based architecture, using static *xy* routing. In [86] the algorithm is extended to route with the objective of balancing network load. In [133, 134, 136] a heuristic improvement method is used. Routing is repeated for pair-wise swaps of IPs in the topology, thereby exploring the design space in search for an efficient mapping. In [136] the algorithm integrates physical planning. In contrast to the aforementioned works, a greedy non-iterative algorithm is presented in [62] that presents a complete interconnect design flow. In this flow, mapping is done based on IP clustering after which paths are selected using static *xy* routing. All the aforementioned works rely on a multi-step approach where mapping is carried out before routing. Routing and mapping objectives do hereby not necessarily coincide. The routing phase must adhere to decisions taken in the mapping phase which invariably limits the routing solution space. In [85, 86, 133, 134, 136], multiple mapping and routing solutions are evaluated iteratively to mitigate the negative effects mapping decisions may have on routing.

Routing objectives are discussed in [66, 195] and implications with common-practice load-balancing solutions are addressed in [119]. Routing algorithms that incorporate temporal guarantees assume static schedules on the application level [65, 81, 136, 157, 188], i.e. known message production and consumption times, thus severely limiting their applicability.

Resource allocation for multiple use-cases is addressed in [137] by generating a synthetic worst-case use-case. The result is one allocation spanning all potential use-cases. This approach is subsumed in [138], where the lack of scalability in the synthetic worst-case solution is addressed by allocating resources on the granularity of aggregated use-cases. Undisrupted services are provided within the constituent use-cases, but do not cover use-case transitions. The reason is that binding of applications to resources is done per use-case [138]. Thus, going from one use-case to another is either done without any change (for the interconnect) or requires global reconfiguration. Not only does this cause a disruption in delivered services, but it leads to unpredictable reconfiguration times as in-flight transactions must be allowed to finish when tearing down communication between IPs [68, 102, 145]. The other alternative, avoiding reconfiguration altogether by allocating a synthetic worst-case use-case covering the requirements of all use-cases [137], leads to a much too costly interconnect design [138]. Moreover, if the platform architecture is given, it may not be possible to find an allocation that meets the worst-case use-case requirements, even though allocations exist for each use-case individually.

In addition to the problem of resource allocation, much work focus on generating, exploring, evaluating, and comparing NoC architectures and instantiations, some targeting FPGAs [11, 58, 129], and others taking a more general approach [17, 23, 62]. Performance verification [57], although with a very limited application model (periodic producers and consumers), is also part of the flow in [62].

In this work, our dimensioning and allocation flow unifies the spatial mapping of IPs, and the spatial and temporal routing of communication. The proposed solution is fundamentally different from [62, 85, 86, 133, 134, 136] in that mapping is no longer done prior to routing but instead during it. Moreover, we consider the communication real-time requirements, and guarantee that constraints on throughput and latency are met, irrespective of how the IPs use the network. This work is, to the best of our knowledge, the first to enable partial reconfiguration of the interconnect with starting and stopping of one application without affecting other applications. Our current allocation flow does, however, not support modes or scenarios within applications, as proposed in [60]. Extending on [62], we offer a complete design flow, going from communication requirements to an instantiated and verified interconnect instance.

Chapter 10
Conclusions and Future Work

The complexity of system verification and integration is exploding due to the growing number of real-time applications integrated on a single chip. In addition, the applications have diverse requirements and behaviours and are started and stopped dynamically at run time. To reduce the design and verification complexity, it is crucial to offer a platform that enables independent implementation, verification and debugging of applications.

In this work we introduce aelite, a composable and predictable on-chip interconnect that eliminates interference between applications without placing any requirements on their behaviour. In the previous chapters, we have shown how to dimension the architecture, how to allocate resources for the applications, how to instantiate the interconnect hardware and software, and how to verify that the result satisfies the application requirements. In this chapter, we revisit the requirements formulated in Chapter 1, and discuss how the different chapters contribute in addressing them (Section 10.1). Finally, we present directions for future work (Section 10.2).

10.1 Conclusions

An overview of the contributions of the individual chapters of this work is provided in Table 10.1. In the table, a positive contribution is indicated with a '+' and a negative contribution with a '−'. In case the specific requirement is not touched upon by the specific chapter, the space is left blank.

Starting with the *scalability* of the proposed interconnect, it is offered at the physical, architectural level and functional level, as demonstrated in Chapter 8. However, the scalability of the design flow is currently limited by the resource allocation and instantiation. As we have seen in the experimental results of Chapter 4, the execution time of the heuristic allocation algorithm grows exponentially in the size of the network topology. Hence, the proposed algorithm is limited to a few hundred IPs and complementary techniques (e.g. partitioning into sub-networks) or completely different algorithms are required to offer scalability also in this part of the design flow. Similarly for the instantiation, as described in Chapter 5, our design flow unconditionally instantiates the entire system (hardware), even if we are only interested

A. Hansson, K. Goossens, *On-Chip Interconnect with Aelite*, Embedded Systems,
DOI 10.1007/978-1-4419-6865-4_10,

Table 10.1 Contributions to the problem statement

	Dimensioning	Allocation	Instantiation	Verification
Scalability	+	−	−	+
Diversity	+	+	+	+
Composability	+	+	−	
Predictability	+	+	+	+
Reconfigurability	+	+	+	
Automation	+	+	+	+

in using the instantiation to design or verify a subset of the applications. To offer scalability, the instantiation should exploit the composability of the interconnect, as discussed in Section 10.2.

Diversity is offered across all parts of the flow. The dimensioning offers diversity in interfaces, programming models and communication paradigms. The allocation offers diversity in the formulation (and interpretation) of requirements, e.g. worst- or average-case. In the instantiation, we offer diversity in accuracy and speed of the simulation models, and in the tooling and implementation of the target platform, e.g. ASIC or FPGA, and different host implementations. Finally, for the verification, our proposed interconnect accommodates probabilistic as well as worst-case analysis, based on either simulation or formal models (or even a combination).

Composability is offered by the architecture as part of the dimensioning, and also in the allocation of resources to individual connections. However, as we have already seen in the discussion of scalability, the composability of our interconnect is not considered in the instantiation. As a result, the composability is negatively affected. To address this issue, the design flow should only instantiate the required subset of the architecture, i.e. the virtual platform of the application in question. The verification relies on, but does not contribute to composability.

Every part of the design flow contributes to the *predictability*. The dimensioning uses modules with predictable behaviour, and the allocation exploits these behaviours when deriving an allocation that satisfies the applications' communication requirements. The instantiation also contributes to the predictability of the interconnect by offering temporal bounds on the reconfiguration operations. Lastly, the verification raises the level of predictability from individual network channels, to end-to-end performance verification at the application level. As we shall see in our discussion of future work, we consider it an important extension to also automate the construction of formal models for the sharing of memory-mapped target ports.

Reconfigurability is an important part of the dimensioning, where a few selected modules enable a host to change the logical topology and the temporal behaviour of the interconnect connections. The dimensioning also contributes by adding a virtual control infrastructure on top of the functional interconnect. The allocation flow has a major contribution to the reconfigurability as it enables dynamic partial reconfiguration. As we have seen in Chapter 2, the granularity of allocations is key in enabling applications to be started and stopped without affecting other applications that keep running. Using the allocations, the run-time library of the instantiation flow contributes to the reconfigurability by ensuring correctness and providing temporal bounds on the configuration operations. As seen in Table 10.1, the verification does not contribute (in either a positive or negative way) to the reconfigurability.

The last requirement, *automation*, is contributed to by all parts of the flow. As we have demonstrated in Chapters 7 and 8, the proposed flow enables us to go from high-level requirements into a system instance within hours, and addresses the need for performance verification for a heterogeneous mix of firm, soft and non real-time applications.

10.2 Future Work

The proposed interconnect serves as a first step towards a composable and predictable system-level design method. However, there are many important extensions and improvements to be made. Our research may be continued in the following directions:

- A first extension to the presented case studies is to also evaluate the performance of the proposed interconnect after layout. Only then is it possible to really assess the contributions of the mesochronous links and the claims on physical scalability.
- Our proposed interconnect does not address deep sub-micron effects, e.g. electromigration, voltage-drop and on-chip variations, that are becoming predominant at 65 nm [112]. Thus, techniques are needed to recover from errors and offer resiliency at the physical as well as the architectural level.
- The power consumption of the interconnect is an important cost aspect not investigated in this work. We do, however, provide a modular interconnect, with ideal boundaries for power and clock domains. Moreover, due to the well-known forwarding delays, datapath self-gating in routers and NIs is a straightforward extension that could be evaluated. The ability to identify quiescence also opens up opportunities for safely gating the power or clock of the NI buffers.
- To increase the benefits of the aforementioned power awareness, the run-time library functions can also be used together with a power management library (when opening and closing connections). Together with hardware support, e.g. clamping of the network link signals, this affords safe power down of parts of the interconnect.
- Currently we assume that the entire interconnect is in one voltage and frequency island, and that there are no run-time variations in supply voltage or frequency. To conserve power, the importance of dynamic voltage and frequency scaling is growing and a possible extension is to split the interconnect into multiple voltage and clock domains, possibly with different word widths.
- Our proposed interconnect currently uses TDM arbiters to achieve composability, also at the shared target ports. As discussed in Chapter 2, this is not a requirement. As long as all arbiters can be characterised as *latency-rate servers* [185] it is possible to eliminate the application interference by *enforcing the worst-case*

bounds. Thus, it is possible to use e.g. rate-controlled priority-based arbiters [3], where unlike TDM it is possible to distinguish between the allocated latency and rate.

- In addition to more elaborate arbitration schemes, we also consider the distinction between inter- and intra-application arbitration an important extension. By separating the two levels, the interconnect is able to offer application composability, and still distribute slack within an application. The additional degree of freedom opens up many new opportunities (and challenges) for the management and allocation of resources.
- Moving from the interconnect architecture to the IPs, we consider the inclusion of caches an important extension. Caches introduce two problems that must be addressed, namely *composable sharing* between applications [125], and *cache coherency*. The interconnect currently supports software-based cache coherency [190] (as it requires no additional functionality besides what is described in Chapter 3), but not hardware-based cache coherency. Moreover, composable sharing of caches is currently not supported.
- Currently. we do not allow memory-mapped initiator ports, as used for example by the processors, to be shared between applications. To offer composable sharing also of these ports and consequently also processors, a mechanism similar to what is proposed in Section 2.4 (i.e. buffers on both sides of the shared resource and flow control and arbitration at the same granularity) is required. In contrast to the hardware solution used for shared target ports, also the operating system (or at least the middleware used for communication) of the processor must be adapted.
- In addition to the architectural extensions, we consider run-time resource allocation, similar to what is proposed in [145], an important extension to our proposed interconnect. Currently we assume that the set of applications is given at compile time. By implementing the allocation algorithms on an embedded processor, it is possible to offer run-time additions of applications. There are, however, many challenges involved in finding and verifying an allocation using very limited computational resources.
- Currently, our design flow (automatically) only generates models for the network channels and protocol shells. Thus, we consider it an important extension to also include formal models of the initiator buses and shared targets. For a given set of (pre-characterised) targets, it would thereby be possible to generate a model of the complete interconnect architecture and resource allocation.

With this rather extensive list of directions for future research, we conclude that the our composable and predictable on-chip interconnect not only serves to make applications first-class citizens in embedded systems, but also opens many new exciting research directions.

Appendix A
Example Specification

The proposed design flow in Fig. 1.8 takes its starting point in a description of the IP architecture and the communication requirements. Here, we show the two example XML files for the system used in Chapter 7.

The architecture specification is shown in Section A.1. The file starts with the architectural constants governing the network behaviour, as introduced in Table 3.1. Note that the slot table size s_{tbl} is not fixed and that the design flow automatically decides on a smallest possible size. After the constants, the clocks in the architecture are listed. The main part of the architecture description is the IPs. Every IP has a unique name and a type that informs the design flow of any specific behaviours, e.g. the host, or implementations, e.g. pearl_ray that refers to a specific VLIW processor. Every IP has a number of ports, also with a unique name (within the block). Each port also has a type and a protocol. The type is either initiator or target, and the protocol is memory-mapped (MMIO_DTL) or streaming (FIFO_AE). For memory-mapped target ports, we also specify the base address and capacity (i.e. address range) for the configuration of the address decoders in the target buses. The last part of the architecture specification is the layout constraints, i.e. the port groups and eligible NIs. Every block that is added and as part of the dimensioning in Chapter 3 uses post-fix naming, e.g. *vliw1_pi_bus* is the (automatically added) target bus connected to the port *pi* on the IP called *vliw1*. Using regular expressions [30], it is thereby easy to group all ports on that bus as *vliw1_pi_bus.** or simply all ports on the IP as *vliw1.**. The specification of eligible NIs also uses regular expressions, but relies on knowledge of the naming scheme used in the topology generation (or a manually specified topology). In this case, we use a 5×1 mesh, as shown in Fig. 2.1, and the expression *NIx(0|2)y0n[0−1]* matches four NIs.

The communication requirements are shown in Section A.2. The description lists the applications, and within each application there are a number of connections. The names of initiator and target ports correspond to the IP and port identifiers in Section A.1. In addition to the start and end point, a connection also specifies a throughput requirement ('bw' in Mbps) and a latency requirement (in ns) for writes and reads, respectively. Note that for connections between two memory-mapped ports, it is also possible to specify a burst size. This is to enable to the design flow to account for the serialisation of request and response messages. As seen in the

A. Hansson, K. Goossens, *On-Chip Interconnect with Aelite*, Embedded Systems, DOI 10.1007/978-1-4419-6865-4, © Springer Science+Business Media, LLC 2011 specification of, e.g., the decoder, it is possible to not specify a latency requirement. The last part of the communication specification is the constraints on how applications are allowed to be combined into use-cases. We rely on regular expressions also here, to formulate the constraints shown in Fig. 1.6a. The use-case constraints for the status application, for example, are specified using negative lookahead [30] to include everything but the initialisation application.

A.1 Architecture

```
<architecture id="fpga">
  <parameter id="clk" type="string" value="fpga_54MHz" />
  <parameter id="maxpcklen" type="int" value="4" unit="flits" />

  <clk id="usb_48MHz" period="20.83" />
  <clk id="fpga_54MHz" period="18.52" />
  <clk id="video_65MHz" period="15.38" />
  <clk id="audio_25MHz" period="39.72" />

  <ip id="host" type="Host">
    <parameter id="clk" type="string" value="usb_48MHz" />
    <port id="pi" type="Initiator" protocol="MMIO_DTL" />
  </ip>

  <ip id="vliw1" type="pearl_ray">
    <port id="pst" type="Target" protocol="FIFO_AE" />
    <port id="psi" type="Initiator" protocol="FIFO_AE" />

    <port id="pt" type="Target" protocol="MMIO_DTL">
      <parameter id="address" type="int" value="0x08000000" />
      <parameter id="capacity" type="int" value="32" unit="kbyte" />
    </port>

    <port id="pi" type="Initiator" protocol="MMIO_DTL" />
  </ip>

  <ip id="vliw2" type="pearl_ray">
    <port id="pst" type="Target" protocol="FIFO_AE" />
    <port id="psi" type="Initiator" protocol="FIFO_AE" />

    <port id="pt" type="Target" protocol="MMIO_DTL">
      <parameter id="address" type="int" value="0x08000000" />
      <parameter id="capacity" type="int" value="32" unit="kbyte" />
    </port>

    <port id="pi" type="Initiator" protocol="MMIO_DTL" />
  </ip>

  <ip id="vliw3" type="pearl_ray">
    <port id="pst" type="Target" protocol="FIFO_AE" />
    <port id="psi" type="Initiator" protocol="FIFO_AE" />
```

```
    <port id="pt" type="Target" protocol="MMIO_DTL">
      <parameter id="address" type="int" value="0x08000000" />
      <parameter id="capacity" type="int" value="32" unit="kbyte" />
    </port>

    <port id="pi" type="Initiator" protocol="MMIO_DTL" />
  </ip>

  <ip id="sram" type="IP">
    <parameter id="clk" type="string" value="video_65MHz" />
    <port id="pt" type="Target" protocol="MMIO_DTL">
      <parameter id="address" type="int" value="0x30000000" />
      <parameter id="capacity" type="int" value="8192" unit="kbyte" />
    </port>
  </ip>

  <ip id="video" type="IP">
    <parameter id="clk" type="string" value="video_65MHz" />
    <port id="pt" type="Target" protocol="MMIO_DTL">
      <parameter id="address" type="int" value="0x38000000" />
      <parameter id="capacity" type="int" value="4096" unit="kbyte" />
    </port>
  </ip>

  <ip id="audio" type="IP">
    <parameter id="clk" type="string" value="audio_25MHz" />
    <port id="pst" type="Target" protocol="FIFO_AE" />
    <port id="psi" type="Initiator" protocol="FIFO_AE" />
  </ip>

  <ip id="peripheral" type="IP">
    <port id="pt" type="Target" protocol="MMIO_DTL">
      <parameter id="address" type="int" value="0x40000000" />
      <parameter id="capacity" type="int" value="16" unit="kbyte" />
    </port>
  </ip>

  <constraint id='host' port='host.*' nis='NIx0y0n0' />
  <constraint id='vliw1' port='vliw1.*' />
  <constraint id='vliw2' port='vliw2.*' />
  <constraint id='vliw3' port='vliw3.*' />
  <constraint id='sram' port='sram.*' nis='NIx(0|2)y0n[0-1]'/>
  <constraint id='video' port='video.*' />
  <constraint id='peripheral' port='peripheral.*' />
</architecture>
```

A.2 Communication

```
<communication>
  <application id='filter'>
    <connection id='0'>
      <initiator ip='vliw1' port='psi'/>
```

```
    <target ip='audio' port='pst'/>
    <write bw='1.5' latency='1000' />
  </connection>
  <connection id='1'>
    <initiator ip='audio' port='psi'/>
    <target ip='vliw1' port='pst'/>
    <write bw='1.5' latency='1000' />
  </connection>
  <connection id='2'>
    <initiator ip='vliw1' port='pi'/>
    <target ip='memory' port='pt'/>
    <write bw='1.5' burstsize='4' latency='500' />
    <read bw='3' burstsize='4' latency='500' />
  </connection>
</application>

<application id='player'>
  <connection id='0'>
    <initiator ip='vliw1' port='psi'/>
    <target ip='audio' port='pst'/>
    <write bw='1.5' latency='1000' />
  </connection>
</application>

<application id='decoder'>
  <connection id='0'>
    <initiator ip='vliw2' port='pi'/>
    <target ip='vliw3' port='pt'/>
    <write bw='8' burstsize='4'/>
  </connection>
  <connection id='1'>
    <initiator ip='vliw2' port='pi'/>
    <target ip='sram' port='pt'/>
    <read bw='1' latency='600' burstsize='4'/>
  </connection>
  <connection id='2'>
    <initiator ip='vliw3' port='pi'/>
    <target ip='vliw2' port='pt'/>
    <write bw='0.1' burstsize='4'/>
  </connection>
  <connection id='3'>
    <initiator ip='vliw3' port='pi'/>
    <target ip='video' port='pt'/>
    <write bw='8' burstsize='4'/>
  </connection>
</application>

<application id='game'>
  <connection id='0'>
    <initiator ip='vliw2' port='pi'/>
    <target ip='peripheral' port='pt'/>
    <write bw='0.1' burstsize='4'/>
  </connection>
```

```
    <connection id='1'>
      <initiator ip='vliw2' port='pi'/>
      <target ip='video' port='pt'/>
      <write bw='8' burstsize='4'/>
    </connection>
  </application>

  <application id='status'>
    <connection id='0'>
      <initiator ip='host' port='pi'/>
      <target ip='peripheral' port='pt'/>
      <write bw='0.1'/>
    </connection>
  </application>

  <application id='init'>
    <connection id='0'>
      <initiator ip='host' port='pi'/>
      <target ip='vliw11' port='pt'/>
      <read bw='0.1' burstsize='64'/>
      <write bw='0.1' burstsize='64'/>
    </connection>
    <connection id='1'>
      <initiator ip='host' port='pi'/>
      <target ip='vliw2' port='pt'/>
      <read bw='0.1' burstsize='64'/>
      <write bw='0.1' burstsize='64'/>
    </connection>
    <connection id='2'>
      <initiator ip='host' port='pi'/>
      <target ip='vliw33' port='pt'/>
      <read bw='0.1' burstsize='64'/>
      <write bw='0.1' burstsize='64'/>
    </connection>
    <connection id='3'>
      <initiator ip='host' port='pi'/>
      <target ip='peripheral' port='pt'/>
      <write bw='0.1' burstsize='64'/>
    </connection>
    <connection id='4'>
      <initiator ip='host' port='pi'/>
      <target ip='sram' port='pt'/>
      <write bw='0.1' burstsize='64'/>
    </connection>

    <tree queue="merged">
       <channel connection="0" direction="request" />
       <channel connection="1" direction="request" />
       <channel connection="2" direction="request" />
       <channel connection="3" direction="request" />
       <channel connection="4" direction="request" />
      </tree>
      <tree queue="merged">
```

```
      <channel connection="0" direction="response" />
      <channel connection="1" direction="response" />
      <channel connection="2" direction="response" />
      <channel connection="3" direction="response" />
      <channel connection="4" direction="response" />
     </tree>
  </application>

  <constraint type="allow" appl="filter"
              with="(status|decoder|init|game)" />
  <constraint type="allow" appl="player"
              with="(status|decoder|init|game)" />
  <constraint type="allow" appl="decoder"
              with="(status|filter|player)" />
  <constraint type="allow" appl="game"
              with="(status|filter|player)" />
  <constraint type="allow" appl="status"
              with="^((?!init).)*$" />
  <constraint type="allow" appl="init"
              with="(player|filter)" />

</communication>
```

References

[1] Abeni L, Buttazzo G (2004) Resource reservation in dynamic real-time systems. Real-Time Systems 27(2):123–167

[2] AHBLite (2001) Multi-Layer AHB, AHB-Lite Product Information. ARM Limited, San Jose, CA

[3] Akesson B, Goossens K, Ringhofer M (2007) Predator: a predictable SDRAM memory controller. In: Proc. CODES+ISSS

[4] Altera (2008) Avalon Interface Specifications. Altera Corporation, San Jose, CA. Available on www.altera.com

[5] Anderson MR (2004) When companies collide: the convergence to consumer electronics. Strategic News Service, Friday Harbor, WA

[6] ARINC653 (1997) ARINC Specification 653. Avionics Application Software Standard Interface

[7] Arteris (2005) A comparison of network-on-chip and busses. White paper

[8] AXI (2003) AMBA AXI Protocol Specification. ARM Limited, San Jose, CA

[9] Azimi M, Cherukuri N, Jayashima D, Kumar A, Kundu P, Park S, Schoinas I, Vaidya A (2007) Integration challenges and tradeoffs for tera-scale architectures. Intel Technology Journal 11(3):173–184

[10] Bambha N, Kianzad V, Khandelia M, Bhattacharyya S (2002) Intermediate representations for design automation of multiprocessor DSP systems. Design Automation for Embedded Systems 7(4):307–323

[11] Bartic T, Desmet D, Mignolet JY, Marescaux T, Verkest D, Vernalde S, Lauwereins R, Miller J, Robert F (2004) Network-on-chip for reconfigurable systems: from high-level design down to implementation. In: Proc. FPL

[12] Beigne E, Clermidy F, Vivet P, Clouard A, Renaudin M (2005) An asynchronous NOC architecture providing low latency service and its multi-level design framework. In: Proc. ASYNC

[13] Beigné E, Clermidy F, Miermont S, Vivet P (2008) Dynamic voltage and frequency scaling architecture for units integration within a GALS NoC. In: Proc. NOCS

[14] Bekooij M, Moreira O, Poplavko P, Mesman B, Pastrnak M, van Meerbergen J (2004) Predictable embedded multiprocessor system design. LNCS 3199:77–91

[15] Bekooij MJG, Smit GJM (2007) Efficient computation of buffer capacities for cyclo-static dataflow graphs. In: Proc. DAC

[16] Bellman R (1957) Dynamic Programming. Princeton University Press, Princeton, NJ

[17] Benini L (2006) Application specific NoC design. In: Proc. DATE

[18] Benini L, de Micheli G (2001) Powering networks on chips. In: Proc. ISSS

[19] Benini L, de Micheli G (2002) Networks on chips: a new SoC paradigm. IEEE Computer 35(1):70–80

[20] van den Berg A, Ren P, Marinissen EJ, Gaydadjiev G, Goossens K (2008) Bandwidth analysis for reusing functional interconnect as test access mechanism. In: Proc. ETS

[21] van Berkel K (2009) Multi-core for mobile phones. In: Proc. DATE
[22] van Berkel K, Heinle F, Meuwissen P, Moerman K, Weiss M (2005) Vector processing as an enabler for software-defined radio in handheld devices. EURASIP Journal on Applied Signal Processing 16:2613–2625
[23] Bertozzi D, Jalabert A, Murali S, Tamhankar R, Stergiou S, Benini L, Micheli GD (2005) NoC synthesis flow for customized domain specific multiprocessor systems-on-chip. IEEE Transactions on Parallel and Distributed Systems 16(2):113–129
[24] Bilsen G, Engels M, Lauwereins R, Peperstraete J (1996) Cyclo-static dataflow. IEEE Transactions on Signal Processing 44(2):397–408
[25] Bjerregaard T, Mahadevan S (2006) A survey of research and practices of network-on-chip. ACM Computing Surveys 38(1):1–51
[26] Bjerregaard T, Sparsø J (2005) A router architecture for connection-oriented service guarantees in the MANGO clockless network-on-chip. In: Proc. DATE
[27] Bjerregaard T, Sparsø J (2005) A scheduling discipline for latency and bandwidth guarantees in asynchronous network-on-chip. In: Proc. ASYNC
[28] Bjerregaard T, Mahadevan S, Grøndahl Olsen R, Sparsø J (2005) An OCP compliant network adapter for GALS-based SoC design using the MANGO network-on-chip. In: Proc. SOC
[29] Bjerregaard T, Stensgaard M, Sparsø J (2007) A scalable, timing-safe, network-on-chip architecture with an integrated clock distribution method. In: Proc. DATE
[30] Boost (2009) Boost c++ libraries. Available from: http://www.boost.org
[31] Bril RJ, Hentschel C, Steffens EF, Gabrani M, van Loo G, Gelissen JH (2001) Multimedia QoS in consumer terminals. In: Proc. SiPS
[32] van de Burgwal MD, Smit GJM, Rauwerda GK, Heysters PM (2006) Hydra: an energy-efficient and reconfigurable network interface. In: Proc. ERSA
[33] Buttazo GC (1977) Hard Real-Time Computing Systems: Predictable Scheduling Algorithms and Applications. Kluwer Publishers, Dordrecth
[34] Cadence SpecMan (2009) Cadence specman. Available from: http://www.cadence.com
[35] Carloni L, McMillan K, Sangiovanni-Vincentelli A (2001) Theory of latency-insensitive design. IEEE Transactions on CAD of Integrated Circuits and Systems 20(9):1059–1076
[36] Chen J, Jone W, Wang J, Lu HI, Chen T (1999) Segmented bus design for low-power systems. IEEE Transactions on VLSI 7(1):25–29
[37] Chen K, Malik S, August D (2001) Retargetable static timing analysis for embedded software. In: Proc. ISSS
[38] Chen S, Nahrstedt K (1998) An overview of quality-of-service routing for the next generation high-speed networks: problems and solutions. IEEE Network 12(6):64–79
[39] Coenen M, Murali S, Rădulescu A, Goossens K, De Micheli G (2006) A buffer-sizing algorithm for networks on chip using TDMA and credit-based end-to-end flow control. In: Proc. CODES+ISSS
[40] Coppola M, Grammatikakis M, Locatelli R, Maruccia G, Pieralisi L (2008) Design of Cost-Efficient Interconnect Processing Units: Spidergon STNoC. CRC Press, Boca Raton, FL
[41] Cummings CE (2002) Simulation and Synthesis Techniques for Asynchronous FIFO Design. Synopsys Users Group, Mountain View, CA
[42] Dales M (2000) SWARM – Software ARM. http://www.cl.cam.ac.uk/~mwd24/phd/swarm.html
[43] Dally WJ, Towles B (2001) Route packets, not wires: on-chip interconnection networks. In: Proc. DAC
[44] Dasdan A (2004) Experimental analysis of the fastest optimum cycle ratio and mean algorithms. ACM Transactions on Design Automation of Electronic Systems 9(4):385–418
[45] Day J, Zimmermann H (1983) The OSI reference model. Proceedings of the IEEE 71(12):1334–1340
[46] Dielissen J, Rădulescu A, Goossens K, Rijpkema E (2003) Concepts and implementation of the Philips network-on-chip. In: IP-Based SOC Design

[47] Diguet J, Evain S, Vaslin R, Gogniat G, Juin E (2007) NOC-centric security of reconfigurable SoC. In: Proc. NOCS
[48] Dijkstra EW (1959) A note on two problems in connection with graphs. Numerische Mathematik 1(5):269–271
[49] DTL (2002) Device Transaction Level (DTL) Protocol Specification. Version 2.2. Philips Semiconductors, Washington, DC
[50] Dutta S, Jensen R, Rieckmann A (2001) Viper: a multiprocessor SOC for advanced set-top box and digital TV systems. IEEE Design and Test of Computers 18(5):21–31
[51] Ernst D (2004) Limits to modularity: a review of the literature and evidence from chip design. Economics Study Area Working Papers 71, East-West Center, Honolulu, HI
[52] Felicijan T (2007) Asynchronous tdma networks on chip. Tech. rep., Royal Philips Electronics
[53] Ford LR, Fulkerson DR (1962) Flows in Networks. Princeton University Press, Princeton, NJ
[54] FSL (2007) Fast Simplex Link (FSL) Bus v2.11a. Xilinx, Inc, San Jose, CA
[55] Gal-On S (2008) Multicore benchmarks help match programming to processor architecture. In: MultiCore Expo
[56] Gangwal O, Rădulescu A, Goossens K, Pestana S, Rijpkema E (2005) Building predictable systems on chip: an analysis of guaranteed communication in the Æthereal network on chip. In: Dynamic and Robust Streaming in and Between Connected Consumer-Electronics Devices, Kluwer, Dordrecht
[57] Gangwal OP, Janssen J, Rathnam S, Bellers E, Duranto M (2003) Understanding video pixel processing applications for flexible implementations. In: Proc. DSD
[58] Genko N, Atienza D, Micheli GD, Mendias J, Hermida R, Catthoor F (2005) A complete network-on-chip emulation framework. In: Proc. DATE
[59] Gharachorloo K, Lenoski D, Laudon J, Gibbons P, Gupta A, Hennessy J (1990) Memory consistency and event ordering in scalable shared-memory multiprocessors. In: Proc. ISCA
[60] Gheorghita SV, Palkovic M, Hamers J, Vandecappelle A, Mamagkakis S, Basten T, Eeckhout L, Corporaal H, Catthoor F, Vandeputte F, De Bosschere K (2009) System-scenario-based design of dynamic embedded systems. ACM Transactions on Design Automation of Electronic Systems 13(1):1–45
[61] Goossens K, Gangwal OP, Röver J, Niranjan AP (2004) Interconnect and memory organization in SOCs for advanced set-top boxes and TV—evolution, analysis, and trends. In: Nurmi J, Tenhunen H, Isoaho J, Jantsch A (eds) Interconnect-Centric Design for Advanced SoC and NoC, Kluwer, Dordrecht Chap 15, pp. 399–423
[62] Goossens K, Dielissen J, Gangwal OP, González Pestana S, Rădulescu A, Rijpkema E (2005) A design flow for application-specific networks on chip with guaranteed performance to accelerate SOC design and verification. In: Proc. DATE
[63] Goossens K, Dielissen J, Rădulescu A (2005) The Æthereal network on chip: concepts, architectures, and implementations. IEEE Design and Test of Computers 22(5):21–31
[64] Graham R (1969) Bounds on multiprocessing timing anomalies. SIAM Journal of Applied Mathematics 17(2):416–429
[65] Guérin R, Orda A (2000) Networks with advance reservations: The routing perspective. In: Proc. INFOCOM
[66] Guérin R, Orda A, Williams D (1997) QoS routing mechanisms and OSPF extensions. In: Proc. GLOBECOM
[67] Halfhill TR (2006) Ambric's new parallel processor. Microprocessor Report
[68] Hansson A, Goossens K (2007) Trade-offs in the configuration of a network on chip for multiple use-cases. In: Proc. NOCS
[69] Hansson A, Goossens K (2009) An on-chip interconnect and protocol stack for multiple communication paradigms and programming models. In: Proc. CODES+ISSS

[70] Hansson A, Goossens K, Rădulescu A (2005) A unified approach to constrained mapping and routing on network-on-chip architectures. In: Proc. CODES+ISSS
[71] Hansson A, Coenen M, Goossens K (2007) Channel trees: reducing latency by sharing time slots in time-multiplexed networks on chip. In: Proc. CODES+ISSS
[72] Hansson A, Coenen M, Goossens K (2007) Undisrupted quality-of-service during reconfiguration of multiple applications in networks on chip. In: Proc. DATE
[73] Hansson A, Goossens K, Rădulescu A (2007) Avoiding message-dependent deadlock in network-based systems on chip. VLSI Design 2007:1–10
[74] Hansson A, Goossens K, Rădulescu A (2007) A unified approach to mapping and routing on a network on chip for both best-effort and guaranteed service traffic. VLSI Design 2007: 1–16
[75] Hansson A, Wiggers M, Moonen A, Goossens K, Bekooij M (2008) Applying dataflow analysis to dimension buffers for guaranteed performance in Networks on Chip. In: Proc. NOCS
[76] Hansson A, Akesson B, van Meerbergen J (2009) Multi-processor programming in the embedded system curriculum. ACM SIGBED Review 6(1)
[77] Hansson A, Goossens K, Bekooij M, Huisken J (2009) Compsoc: a template for composable and predictable multi-processor system on chips. ACM Transactions on Design Automation of Electronic Systems 14(1):1–24
[78] Hansson A, Subburaman M, Goossens K (2009) Aelite: a flit-synchronous network on chip with composable and predictable services. In: Proc. DATE
[79] Hansson A, Wiggers M, Moonen A, Goossens K, Bekooij M (2009) Enabling application-level performance guarantees in network-based systems on chip by applying dataflow analysis. IET Computers and Design Techniques
[80] Henkel J (2003) Closing the SoC design gap. Computer 36(9):119–121
[81] Ho WH, Pinkston TM (2003) A methodology for designing efficient on-chip interconnects on well-behaved communication patterns. In: Proc. HPCA
[82] Holzenspies P, Hurink J, Kuper J, Smit G (2008) Run-time spatial mapping of streaming applications to a heterogeneous multi-processor system-on-chip MPSoC. In: Proc. DATE
[83] Hoskote Y, Vangal S, Singh A, Borkar N, Borkar S (2007) A 5-GHz mesh interconnect for a teraflops processor. IEEE Micro 27(5):51–61
[84] Hu J, Marculescu R (2004) Application-specific buffer space allocation for networks-on-chip router design. In: Proc. ICCAD
[85] Hu J, Mărculescu R (2003) Energy-aware mapping for tile-based NoC architectures under performance constraints. In: Proc. ASP-DAC
[86] Hu J, Mărculescu R (2003) Exploiting the routing flexibility for energy/performance aware mapping of regular NoC architectures. In: Proc. DATE
[87] Ilitzky DA, Hoffman JD, Chun A, Esparza BP (2007) Architecture of the scalable communications core's network on chip. IEEE Micro 27(5):62–74
[88] ITRS (2007) International technology roadmap for semiconductors. System Drivers
[89] ITRS (2007) International technology roadmap for semiconductors. Design
[90] Jantsch A (2006) Models of computation for networks on chip. In: Proc. ACSD
[91] Jantsch A, Tenhunen H (2003) Will networks on chip close the productivity gap? In: Networks on Chip, Kluwer Academic Publishers, Dordrecht, pp. 3–18
[92] Jerraya A, Bouchhima A, Pétrot F (2006) Programming models and HW-SW interfaces abstraction for multi-processor SoC. In: Proc. DAC
[93] Kang J, Henriksson T, van der Wolf P (2005) An interface for the design and implementation of dynamic applications on multi-processor architecture. In: Proc. ESTImedia
[94] Kar K, Kodialam M, Lakshman TV (2000) Minimum interference routing of bandwidth guaranteed tunnels with MPLS traffic engineering applications. IEEE Journal on Selected Areas in Communications 18(12):2566–2579
[95] Kavaldjiev N (2006) A run-time reconfigurable network-on-chip for streaming DSP applications. PhD thesis, University of Twente

[96] Keutzer K, Malik S, Newton AR, Rabaey JM, Sangiovanni-Vincentelli A (2000) System-level design: orthogonalization of concerns and platform-based design. IEEE Transactions on CAD of Integrated Circuits and Systems 19(12):1523–1543
[97] Kollig P, Osborne C, Henriksson T (2009) Heterogeneous multi-core platform for consumer multimedia applications. In: Proc. DATE
[98] Kopetz H (1997) Real-Time Systems: Design Principles for Distributed Embedded Applications. Kluwer Academic Publishers, Dordrecht
[99] Kopetz H, Bauer G (2003) The time-triggered architecture. Proceedings of the IEEE 91(1):112–126
[100] Kopetz H, Obermaisser R, Salloum CE, Huber B (2007) Automotive software development for a multi-core system-on-a-chip. In: Proc. SEAS
[101] Kopetz H, El Salloum C, Huber B, Obermaisser R, Paukovits C (2008) Composability in the time-triggered system-on-chip architecture. In: Proc. SOCC
[102] Kramer J, Magee J (1990) The evolving philosophers problem: dynamic change management. IEEE Transactions on Software Engineering 16(11):1293–1306,
[103] Krstić M, Grass E, Gürkaynak F, Vivet P (2007) Globally asynchronous, locally synchronous circuits: overview and outlook. IEEE Design and Test of Computers 24(5):430–441
[104] Kumar A, Mesman B, Theelen B, Corporaal H, Ha Y (2008) Analyzing composability of applications on MPSoC platforms. Journal of Systems Architecture 54(3–4):369–383
[105] Laffely A (2003) An interconnect-centric approach for adapting voltage and frequency in heterogeneous system-on-a-chip. PhD thesis, University of Massachusetts Amherst
[106] kyung Lee Y, Lee S (2001) Path selection algorithms for real-time communication. In: Proc. ICPADS
[107] Leijten J, van Meerbergen J, Timmer A, Jess J (2000) Prophid: a platform-based design method. Journal of Design Automation for Embedded Systems 6(1):5–37
[108] Liang J, Swaminathan S, Tessier R (2000) aSOC: A scalable, single-chip communications architecture. In: Proc. PACT
[109] Lickly B, Liu I, Kim S, D Patel H, Edwards SA, Lee EA (2008) Predictable programming on a precision timed architecture. In: Proc. CASES
[110] Liu G, Ramakrishnan KG (2001) A*Prune: an algorithm for finding K shortest paths subject to multiple constraints. In: Proc. INFOCOM
[111] Lu Z, Haukilahti R (2003) NOC application programming interfaces: high level communication primitives and operating system services for power management. In: Networks on Chip, Kluwer Academic Publishers, Dordrecht
[112] Magarshack P, Paulin PG (2003) System-on-chip beyond the nanometer wall. In: Proc. DAC
[113] Mangano D, Locatelli R, Scandurra A, Pistritto C, Coppola M, Fanucci L, Vitullo F, Zandri D (2006) Skew insensitive physical links for network on chip. In: Proc. NANONET
[114] Marescaux T, Corporaal H (2007) Introducing the SuperGT network-on-chip. In: Proc. DAC
[115] Marescaux T, Mignolet J, Bartic A, Moffat W, Verkest D, Vernalde S, Lauwereins R (2003) Networks on chip as hardware components of an OS for reconfigurable systems. In: Proc. FPL
[116] Martin A, Nystrom M (2006) Asynchronous techniques for system-on-chip design. Proceedings of the IEEE 94(6):1089–1120
[117] Martin G (2006) Overview of the MPSoC design challenge. In: Proc. DAC
[118] Martin P (2005) Design of a virtual component neutral network-on-chip transaction layer. In: Proc. DATE
[119] Matta I, Bestavros A (1998) A load profiling approach to routing guaranteed bandwidth flows. In: Proc. INFOCOM
[120] Mercer CW, Savage S, Tokuda H (1994) Processor capacity reserves: operating system support for multimedia systems. In: Proc. ICMCS
[121] Messerschmitt D (1990) Synchronization in digital system design. IEEE Journal on Selected Areas in Communication 8(8):1404–1419

[122] Millberg M, Nilsson E, Thid R, Jantsch A (2004) Guaranteed bandwidth using looped containers in temporally disjoint networks within the Nostrum network on chip. In: Proc. DATE
[123] Millberg M, Nilsson E, Thid R, Kumar S, Jantsch A (2004) The nostrum backbone – a communication protocol stack for networks on chip. In: Proc. VLSID
[124] Miro Panades I, Greiner A (2007) Bi-synchronous FIFO for synchronous circuit communication well suited for network-on-chip in GALS architectures. In: Proc. NOCS
[125] Molnos A, Heijligers M, Cotofana S (2008) Compositional, dynamic cache management for embedded chip multiprocessors. In: Proc. DATE
[126] Moonen A (2004) Modelling and simulation of guaranteed throughput channels of a hard real-time multiprocessor system. Master's thesis, Eindhoven University of Technology
[127] Moonen A, Bekooij M, van den Berg R, van Meerbergen J (2007) Practical and accurate throughput analysis with the cyclo static data flow model. In: Proc. MASCOTS
[128] Moonen A et al. (2005) A multi-core architecture for in-car digital entertainment. In: Proc. GSPX
[129] Moraes F, Calazans N, Mello A, Möller L, Ost L (2004) HERMES: an infrastructure for low area overhead packet-switching networks on chip. Integration VLSI Journal 38(1):69–93
[130] Moreira O, Bekooij M (2007) Self-timed scheduling analysis for real-time applications. EURASIP Journal on Advances in Signal Processing pp. 1–14
[131] Moreira O, Valente F, Bekooij M (2007) Scheduling multiple independent hard-real-time jobs on a heterogeneous multiprocessor. In: Proc. EMSOFT
[132] MTL (2002) Memory Transaction Level (MTL) Protocol Specification. Philips Semiconductors, Washington, DC
[133] Murali S, de Micheli G (2004) Bandwidth-constrained mapping of cores onto NoC architectures. In: Proc. DATE
[134] Murali S, de Micheli G (2004) SUNMAP: a tool for automatic topology selection and generation for NoCs. In: Proc. DAC
[135] Murali S, De Micheli G (2005) An application-specific design methodology for STbus crossbar generation. In: Proc. DATE
[136] Murali S, Benini L, de Micheli G (2005) Mapping and physical planning of networks on chip architectures with quality of service guarantees. In: Proc. ASP-DAC
[137] Murali S, Coenen M, Rădulescu A, Goossens K, De Micheli G (2006) Mapping and configuration methods for multi-use-case networks on chips. In: Proc. ASP-DAC
[138] Murali S, Coenen M, Rădulescu A, Goossens K, De Micheli G (2006) A methodology for mapping multiple use-cases on to networks on chip. In: Proc. DATE
[139] Muttersbach J, Villiger T, Fichtner W (2000) Practical design of globally-asynchronous locally-synchronous systems. In: Proc. ASYNC
[140] Nachtergaele L, Catthoor F, Balasa F, Franssen F, De Greef E, Samsom H, De Man H (1995) Optimization of memory organization and hierarchy for decreased size and power in video and image processing systems. In: Proc. MTDT
[141] Nachtergaele L, Moolenaar D, Vanhoof B, Catthoor F, De Man H (1998) System-level power optimization of video codecs on embedded cores: a systematic approach. Journal of VLSI Signal Processing 18(12):89–109
[142] Nelson A (2009) Conservative application-level performance analysis through simulation of a multiprocessor system on chip. Master's thesis, Eindhoven University of Technology
[143] Nesbit K, Moreto M, Cazorla F, Ramirez A, Valero M, Smith J (2008) Multicore resource management. IEEE Micro 28(3):6–16
[144] Nieuwland A, Kang J, Gangwal O, Sethuraman R, Busá N, Goossens K, Peset Llopis R, Lippens P (2002) C-HEAP: a heterogeneous multi-processor architecture template and scalable and flexible protocol for the design of embedded signal processing systems. Design Automation for Embedded Systems 7(3):233–270
[145] Nollet V, Marescaux T, Avasare P, Mignolet JY (2005) Centralized run-time resource management in a network-on-chip containing reconfigurable hardware tiles. In: Proc. DATE

[146] Obermaisser R (2007) Integrating automotive applications using overlay networks on top of a time-triggered protocol. LNCS 4888:187–206
[147] OCP (2007) OCP Specification 2.2. OCP International Partnership
[148] Ogras UY, Hu J, Marculescu R (2005) Key research problems in NoC design: a holistic perspective. In: Proc. CODES+ISSS
[149] Ohbayashi S et al. (2007) A 65-nm SoC embedded 6T-SRAM designed for manufacturability with read and write operation stabilizing circuits. IEEE Journal of Solid-State Circuits 42(4):820–829
[150] Ost L, Mello A, Palma J, Moraes F, Calazans N (2005) MAIA: a framework for networks on chip generation and verification. In: Proc. ASP-DAC
[151] Osteen R, Tou J (1973) A clique-detection algorithm based on neighborhoods in graphs. International Journal of Parallel Programming 2(4):257–268
[152] Owens J, Dally W, Ho R, Jayasimha D, Keckler S, Peh LS (2007) Research challenges for on-chip interconnection networks. IEEE Micro 27(5):96–108
[153] Palesi M, Holsmark R, Kumar S, Catania V (2009) Application specific routing algorithms for networks on chip. IEEE Transactions on Parallel and Distributed Systems 20(3):316–330
[154] Panades I, Greiner A, Sheibanyrad A (2006) A low cost network-on-chip with guaranteed service well suited to the GALS approach. In: Proc. NANONET
[155] Pardalos PM, Rendl F, Wolkowicz H (1994) The quadratic assignment problem: a survey and recent developments. In: Quadratic Assignment and Related Problems. DIMACS Series in Discrete Mathematics and Theoretical Computer Science, vol 16, American Mathematical Society, Providence, RI
[156] Paukovits C (2008) The time-triggered system-on-chip architecture. PhD thesis, Technische Universität Wien, Institut für Technische Informatik
[157] Paukovits C, Kopetz H (2008) Concepts of switching in the time-triggered network-on-chip. In: Proc. RTCSA
[158] Peeters A, van Berkel K (2001) Synchronous handshake circuits. In: Proc. ASYNC
[159] PIBus (1994) PI-Bus Standard OMI 324. Siemens AG, ver. 0.3d edn
[160] PLB (2003) Processor Local Bus (PLB) v3.4. Xilinx Inc, San Jose, CA
[161] Poplavko P, Basten T, Bekooij M, van Meerbergen J, Mesman B (2003) Task-level timing models for guaranteed performance in multiprocessor networks-on-chip. In: Proc. CASES
[162] Pullini A, Angiolini F, Murali S, Atienza D, De Micheli G, Benini L (2007) Bringing NoCs to 65 nm. IEEE Micro 27(5):75–85
[163] Rajkumar R, Juvva K, Molano A, Oikawa S (1998) Resource kernels: a resource-centric approach to real-time systems. In: Proc. MMCN
[164] Rijpkema E, Goossens K, Rădulescu A, Dielissen J, van Meerbergen J, Wielage P, Waterlander E (2003) Trade offs in the design of a router with both guaranteed and best-effort services for networks on chip. IEE Proceedings Computers and Digital Techniques 150(5):294–302
[165] Rostislav D, Vishnyakov V, Friedman E, Ginosar R (2005) An asynchronous router for multiple service levels networks on chip. In: Proc. ASYNC
[166] Rowen C, Leibson S (2004) Engineering the Complex SOC: Fast, Flexible Design with Configurable Processors. Prentice Hall PTR, Upper Saddle River, NJ
[167] Rădulescu A, Goossens K (2004) Communication services for network on silicon. In: Domain-Specific Processors: Systems, Architectures, Modeling, and Simulation, Marcel Dekker, New York, NY
[168] Rădulescu A, Dielissen J, González Pestana S, Gangwal OP, Rijpkema E, Wielage P, Goossens K (2005) An efficient on-chip network interface offering guaranteed services, shared-memory abstraction, and flexible network programming. IEEE Transactions on CAD of Integrated Circuits and Systems 24(1):4–17
[169] Rumpler B (2006) Complexity management for composable real-time systems. In: Proc. ISORC, IEEE Computer Society, Washington, DC
[170] Rutten M, Pol EJ, van Eijndhoven J, Walters K, Essink G (2005) Dynamic reconfiguration of streaming graphs on a heterogeneous multiprocessor architecture. IS&T/SPIE Electron Imag 5683

[171] Saha D, Mukherjee S, Tripathi S (1994) Multi-rate traffic shaping and end-to-end performance guarantees in ATM networks. In: Proc. ICNP
[172] Sasaki H (1996) Multimedia complex on a chip. In: Proc. ISSCC
[173] Sgroi M, Sheets M, Mihal A, Keutzer K, Malik S, Rabaey J, Sangiovanni-Vincentelli A (2001) Addressing the system-on-a-chip interconnect woes through communication-based design. In: Proc. DAC
[174] Shoemaker D (1997) An optimized hardware architecture and communication protocol for scheduled communication. PhD thesis, Massachusetts Institute of Technology
[175] Silicon Hive (2007) Silicon hive. Available from: http://www.siliconhive.com
[176] Smit LT, Smit GJM, Hurink JL, Broersma H, Paulusma D, Wolkotte PT (2004) Run-time mapping of applications to a heterogeneous reconfigurable tiled system on chip architecture. In: Proc. FPT
[177] Smith B (2008) ARM and Intel battle over the mobile chip's future. Computer 41(5):15–18
[178] Song YH, Pinkston TM (2000) On message-dependent deadlocks in multiprocessor/multicomputer systems. In: Proc. HiPC
[179] SonicsMX (2005) SonicsMX Datasheet. Sonics, Inc. Available on www.sonicsinc.com
[180] Soudris D, Zervas ND, Argyriou A, Dasygenis M, Tatas K, Goutis C, Thanailakis A (2000) Data-reuse and parallel embedded architectures for low-power, real-time multimedia applications. In: Proc. PATMOS
[181] Srinivasan K, Chatha KS, Konjevod G (2005) An automated technique for topology and route generation of application specific on-chip interconnection networks. In: Proc. ICCAD
[182] Sriram S, Bhattacharyya S (2000) Embedded Multiprocessors: Scheduling and Synchronization. CRC Press, Boca Raton, FL
[183] Steenhof F, Duque H, Nilsson B, Goossens K, Peset Llopis R (2006) Networks on chips for high-end consumer-electronics TV system architectures. In: Proc. DATE
[184] Stergiou S, Angiolini F, Carta S, Raffo L, Bertozzi D, de Micheli G (2005) ×pipes Lite: a synthesis oriented design library for networks on chips. In: Proc. DATE
[185] Stiliadis D, Varma A (1998) Latency-rate servers: a general model for analysis of traffic scheduling algorithms. IEEE/ACM Transactions on Networking 6(5):611–624
[186] Stoica I, Zhang H (1999) Providing guaranteed services without per flow management. In: Proc. SIGCOMM
[187] Stuijk S, Basten T, Mesman B, Geilen M (2005) Predictable embedding of large data structures in multiprocessor networks-on-chip. In: Proc. DSD
[188] Stuijk S, Basten T, Geilen M, Ghamarian A, Theelen B (2008) Resource-efficient routing and scheduling of time-constrained streaming communication on networks-on-chip. Journal of Systems Architecture 54(3–4):411–426
[189] TechInsights (2008) Embedded Market Study, TechInsights, Ottawa ON
[190] van den Brand J, Bekooij M (2007) Streaming consistency: a model for efficient MPSoC design. In: Proc. DSD
[191] Vercauteren S, Lin B, De Man H (1996) Constructing application-specific heterogeneous embedded architectures from custom HW/SW applications. In: Proc. DAC
[192] Vermeulen B, Goossens K, Umrani S (2008) Debugging distributed-shared-memory communication at multiple granularities in networks on chip. In: Proc. NOCS
[193] Weber WD, Chou J, Swarbrick I, Wingard D (2005) A quality-of-service mechanism for interconnection networks in system-on-chips. In: Proc. DATE
[194] Wentzlaff D, Griffin P, Hoffmann H, Bao L, Edwards B, Ramey C, Mattina M, Miao CC, Brown JF, Agarwal A (2007) On-chip interconnection architecture of the tile processor. IEEE Micro 27(5):15–31
[195] Widyono R (1994) The design and evaluation of routing algorithms for real-time channels. Tech. Rep. TR-94-024, University of California at Berkeley & Int'l Comp. Sci. Inst.
[196] Wielage P, Marinissen E, Altheimer M, Wouters C (2007) Design and DfT of a high-speed area-efficient embedded asynchronous FIFO. In: Proc. DATE

[197] Wiggers M, Bekooij M, Jansen P, Smit G (2007) Efficient computation of buffer capacities for cyclo-static real-time systems with back-pressure. In: Proc. RTAS

[198] Wiggers M, Bekooij M, Smit G (2007) Modelling run-time arbitration by latency-rate servers in dataflow graphs. In: Proc. SCOPES

[199] Wiggers MH, Bekooij MJ, Smit GJ (2008) Buffer capacity computation for throughput constrained streaming applications with data-dependent inter-task communication. In: Proc. RTAS

[200] Wiklund D, Liu D (2003) SoCBUS: switched network on chip for hard real time embedded systems. In: Proc. IPDPS

[201] Wingard D (2004) Socket-based design using decoupled interconnects. In: Interconnect-Centric design for SoC and NoC, Kluwer, Dordrecht

[202] Wingard D, Kurosawa A (1998) Integration architecture for system-on-a-chip design. In: Proc. CICC

[203] Wolkotte P, Smit G, Rauwerda G, Smit L (2005) An energy-efficient reconfigurable circuit-switched network-on-chip. In: Proc. IPDPS

[204] Wüst C, Steffens L, Verhaegh W, Bril R, Hentschel C (2005) QoS control strategies for high-quality video processing. Real-Time Systems 30(1):7–29

[205] Zhang H (1995) Service disciplines for guaranteed performance service in packet-switching networks. Proceedings of the IEEE 83(10):1374–1396

Glossary

Acronyms and Abbreviations

ADC	Analog to digital converter
ANSI	American National Standards Institute
API	Application program interface
ASIC	Application-specific integrated circuit
AXI	Advanced eXtensible interface [8]
CC	Colour conversion
CSDF	Cyclo-static dataflow
DAC	Digital to analog converter
DSP	Digital signal processor
DTL	Device transaction level [49]
FPGA	Field-programmable gate array
FSM	Finite state machine
GALS	Globally asynchronous locally synchronous
HDL	Hardware description language
IDCT	Inverse discrete cosine transform
IP	Intellectual property
ISV	Independent software vendor
ITRS	International technology roadmap for semiconductors
JPEG	Joint photographic experts group
MoC	Model of computation
MPSoC	Multi-processor system on chip
MTL	Memory transaction level [132]
NI	Network interface
NoC	Network on chip
NRE	Non-recurring engineering
OCP	Open core protocol [147]
OSI	Open system interconnection
QAP	Quadratic assignment problem
RTL	Register transfer level
RTTTL	Ring tone text transfer language

SoC	System on chip
Tcl	Tool command language
TDM	Time-division multiplexing
TTM	Time to market
VC	Virtual channel
VLD	Variable length decoder
VLIW	Very long instruction word
VT	Voltage threshold
XML	eXtensible markup language

Symbols and Notations

The list of symbols and notations is split into six categories. The first category contains the symbols and notations used for architectural constants. The second category is related to the input specification of application requirements. The third category covers the description of network topologies. This is followed by symbols and notations used to describe resource allocation in the fourth category. The fifth category contains the symbols and notations used in to capture the intermediate residual resources during the allocation The sixth and last category pertains to the symbols used to describe the temporal behaviour in the channel model.

Architectural Constants

s_{flit}	flit size (words)
s_{tbl}	slot table size (flits)
s_{hdr}	packet header size (words)
s_{pkt}	maximum packet size (flits)
s_{crd}	maximum credits per header (words)
p_n	slot table period (cycles)
$\theta_{\text{p,NI}}$	NI (de)packetisation latency (cycles)
$\theta_{\text{d,NI}}$	NI data pipelining latency (cycles)
$\theta_{\text{c,NI}}$	NI credit pipelining latency (cycles)

Application Requirements

a	application
P_a	set of ports in application a
C_a	set of channels in application a
$\rho(c)$	minimum throughput of channel c
$\theta(c)$	maximum latency of channel c
$\text{src}(c)$	source port of channel c

dst(c)	destination port of channel c
$[c]$	equivalence class (tree) of a channel c
A	set of applications
P	complete set of ports
C	complete set of channels
u	use-case
U	set of use-cases
U_a	set of use-cases with application a
U_c	set of use-cases with a channel c
q	mapping group
Q	set of mapping groups
$[p]$	equivalence class (group) of a port p

Network Topology

N	set of network nodes
L	set of network links
N_{r}	set of routers and link-pipeline stage nodes
N_{n}	set of network interface nodes
N_{q}	set of mapping-group nodes
src(l)	source node of a link l
dst(l)	destination node of a link l
N_{m}	set of mapping nodes
$\epsilon(q)$	set of eligible network interfaces for a group q

Resource Allocation

$\mathrm{map}_i(q)$	mapping of port group q
P'_i	set of mapped ports
$\mathrm{alc}_i(c)$	allocation for channel c
ϕ	network path
T	set of time slots
S_{tbl}	the set of natural numbers smaller than s_{tbl}

Residual Resources

t	slot table
$\sigma(c, t)$	set of slots available for channel c in table t
$t_i(u, l)$	residual slot table for link l in use-case u
$cap_i(u, l)$	residual capacity of link l in use-case u
$use_i(u, l)$	set of channels using link l in use-case u
$t_i(u, \phi)$	slot table for path ϕ in use-case u

$t_i(c, \phi)$	aggregate slot table for channel c after traversing ϕ
$\mathrm{cap}_i(c, l)$	aggregate capacity of link l for channel c
$\mathrm{use}_i(c, l)$	aggregate set set of channels using a link l

Channel Model

$d_\mathrm{d}(T)$	upper bound on latency for data subjected to T
$\hat{h}(T)$	upper bound on number of headers in T
$\theta_\mathrm{p}(\phi)$	path latency
$\check{h}(T)$	lower bound on number of headers in T
$d_\mathrm{c}(T)$	upper bound on latency for credits subjected to T

Index